SUSTAINABILITY AND QUALITY OF LIFE

Edited by

Jack Lee
National Central University, *Taiwan*

SUSTAINABILITY AND QUALITY OF LIFE

Edited by

Jack Lee

EDITOR-IN-CHIEF: CHARLES TANDY

Ria University Press

www.ria.edu/rup

2010

Printed in the United States of America

Ria University Press   **Palo Alto, California**

SUSTAINABILITY AND QUALITY OF LIFE

Edited by

Jack Lee

FIRST PUBLISHED IN 2010

PUBLISHED BY
Ria University Press
PO Box 20170 at Stanford
Palo Alto, California 94309 USA

www.ria.edu/rup

Distributed by Ingram
Available from most bookstores and all Espresso Book Machines

EDITOR-IN-CHIEF: CHARLES TANDY

Copyright © 2010 by Charles Tandy

Paperback/Softcover ISBN-13: 978-0-9743472-1-9

I dedicate this book to

William Grey
The University of Queensland

Contents

    Preface   ix

    Contributors   xi

1. Metaphysics of Sustainability: Kant's Categorical Imperative
   Martin Schönfeld   1

2. Sustainability: A Personal Account
   J. Baird Callicott   19

3. Intrinsic Value and Respect for the Natural Environment
   Jack Lee   35

4. The Land Ethic and Gleason's Individualistic Concept of Plant Association
   Allen Yu   51

5. Environmental Ethics and Bioethics: Anthropocentrism, Ideological Convergence, and Socio-Political Disposition
   Edmund U. H. Sim   77

6. Sustainable Development vs. Sustainable Biosphere
   Holmes Rolston, III   91

7. The Possibility of a Global Environmental Ethics: A Confucian Proposal
   Shui Chuen Lee   103

8. Confucian Filial Piety and Environmental Sustainability
   A. T. Nuyen   119

9. Toward An Ethical Climate Regime
   Po-Keung Ip   137

10. Climate Change and Obligations to the Future
    William Grey   157

11. Environmental Ethics in an Omniverse Environment: From Terrestrial Chauvinism to Golden Rule
Charles Tandy 171

12. The Non-human Natural World, Indigenous Peoples, and Late-modern Capitalism
康柏 **Mac Kang Bai (Campbell)** 217

13. Indigenous People's Hunting Issues and Environmental Ethics: A Contextual Observation in Taiwan
Yih-Ren Lin 255

14. Taiwan's Reform of Energy Law and Policy for Mitigation of Climate Change
Jui-Chu Lin, Tsung-Tang Lee 295

15. How Students Conceptualize the Environment: Implications for Science and Environmental Education
Shiang-Yao Liu 313

Preface

When I did my PhD in Australia, I traveled around the big country. I saw a great variety of scenes: extensive pasture land; sheep and cows wandering freely. I saw wilderness, wetlands, deserts and long white beaches. I said to myself, "Wow, this is good!" Back in Taiwan after finishing my PhD, I did a lot of bushwalking. I'm stunned by the magnificent Taroko National Park. We have spectacular mountains. It's amazing that on our island we have so many bountiful places. I say to myself, "This is also good!" My intuition is that we should preserve and protect these beautiful places. As a philosopher I decided to act on that intuition by working on reasons why we should preserve and protect the natural world.

My attention was also drawn to emergent environmental problems such as climate change, the hole in the ozone layer, species extinction...etc. What should we do about all of this? Can we just give up modern technology and most of our present economic activities, to go back to our ancestors' living styles around the time of Confucius? No, I don't think this is a good way. Some scholars suggest we should live a *sustainable* life. But what is "sustainability"? We need to think about sustainability deeply and broadly. In this anthology, I provide a forum for doing this. Philosophy (e.g., ontology and ethics) is very important. We also need to bring together natural sciences, education, sociology, law and other disciplines. Environmental issues command a multi-disciplinary approach. I am especially proud to include an environmental philosophy paper written by the brilliant young Taiwanese mountaineer Allen Yu before he died, of pollution-related leukemia. I hope this anthology can help us to re-think our engagement with the natural world.

I am very grateful for the support from the Graduate Institute of Philosophy at National Central University and particularly from Professor Jenn-Bang Shiau's project. I also thank photographer Andy Y.C. Lu for the vivid photograph of the owl which I put on the cover. I owe a debt to my graduate students Wei-Lun Chung and Meng-Jie Wu for help with proofreading and editing. I am grateful, finally, to Mac Campbell and Laurence Brown for assistance of many kinds and for general help with this book.

J.L.

Contributors

J. Baird Callicott is Professor of Philosophy and Religion at University of North Texas.

Mac Campbell is an Ecocritic in Philosophy at The University of Queensland.

William Grey is Associate Professor of Philosophy at The University of Queensland.

Po-Keung IP is Professor of Philosophy at National Central University.

Jack Lee is Assistant Professor of Philosophy at National Central University.

Shui Chuen Lee is Professor of Philosophy at National Central University.

Tsung-Tang Lee obtained his LLM at the University of Edinburgh

Jui-Chu Lin is Professor of Law at National Taiwan University of Science and Technology.

Yih-Ren Lin is Associate Professor of Ecology at Providence University.

Shiang-Yao Liu is Associate Professor of Science Education at National Taiwan Normal University.

Anh Tuan Nuyen is Associate Professor of Philosophy at National University of Singapore.

Holmes Rolston, III is Professor of Philosophy at Colorado State University.

Martin Schonfeld is Professor of Philosophy at University of South Florida.

Edmund Ui-Hang Sim is Associate Professor of Molecular Biology at University of Malaysia.

Charles Tandy is Associate Professor of Philosophy at Fooyin University.

The late Allen Yu was a Ph.D. candidate in environmental philosophy at The University of Queensland.

CHAPTER ONE

Metaphysics of Sustainability
Kant's Categorical Imperative

Martin Schönfeld

§ 1

The Paradigm Shift

Sustainability is new. There are older people alive who remember a time when it wasn't taught in school, and when it wasn't a field of study in its own right. In international policy sustainability showed up on the radar screen with the Brundtland Commission Report *Our Common Future* to the UN in 1987.[1] Before reaching this stage, the term had been in use by experts already for several years. The annuals of the Washington, DC, based World Watch Institute on environmental challenges, which have been published since 1983, are self-described reports on progress toward a sustainable society. Such progress has multiple dimensions, as the diverse topics of the annual reports illustrate: the rise of the East, the urban future, sustainable economy, the warming world, and the evolution from consumerism to sustainability.[2] The term is now best known through the *Common Future* definition, according to which sustainable development "is development that meets the needs of the present without compromising the ability of future generations to meet their own needs (ch. 2, #1; cf. also ch. 1, # 49).

The conceptual career of the term is startling. As an abundance of publications illustrates today, sustainability has moved from a specialized environmental policy niche to the interdisciplinary core. Sustainability is now as central in policy as climate change is in the sciences. The meteoric rise of both subjects—sustainability and climate—to the top of interdisciplinary concern indicates a paradigm shift.

Before this shift, and throughout the twentieth century, environmental problems were forming an ever longer list of issues whose interconnectedness was tenuous at best. The issues were troubling but specific, and they were about wildly different things. Some involved endangered species, others the tragedy of the commons, yet other resource exploitation, and still others degrading systems. All of them had something to do with human presence, or collective conduct, or willful activity. Often the events responsible for the issue at stake amounted to point-pollution; industrial smokestacks streaming out soot, or a sewage pipe spilling waste. Acts of destructions used to be heterogeneous but well-circumscribed: clear-cutting a hillside, dynamite-fishing a bay, or draining and paving a wetland. The normative analysis of these actions was often straightforward. The actions could be judged to be violent, irresponsible, selfish, or lacking in traditional virtue—they were destructive, and that's that.

With the paradigm shift, the list of disparate issues folded into a system of interwoven concerns. At stake used to be such things as the protection of rare orchids, vanishing amphibians, shrinking jungles, or stressed coral reefs. At stake now is the continued balance of the Earth system: the complex interplay of the four planetary regions (atmosphere, hydrosphere, biosphere, and lithosphere) through which energy and matter circulate, and whose equilibrium makes us possible.

The shift happened, essentially, because of economic and demographic growth. Consumption, waste, emissions, and population are now at levels in excess of planetary carrying capacity. The weight of overconsumption and overpopulation warps the system that supports it, tilting it out of balance. Emissions are cleaner than they used to be, but sources have multiplied. The random smokestack despoiling a landscape or choking a neighborhood has yielded to countless small exhaust pipes of cars and trucks, fuming away in any landscape, in any neighborhood, anywhere and anytime on the planet.

Point pollution has dissipated into diffuse pollution. There's something genuinely strange about this aspect of the shift.

Pollutants that used to matter were notorious poisons, such as dioxin and DDT, or heavy metals such as mercury, plutonium, or lead; or silicate minerals such as asbestos. Now the decisive pollutants aren't even pollutants in the sense of something toxic or dirty anymore. They can be plain plastic—that re-emerges in the Pacific Trash Vortex, the garbage patch the size of Texas swirling in the north central Pacific. Or they could be greenhouse gases (GHG), such as CO_2, a perfectly normal, natural and clean gas— that accumulates to a level that may push climate to new state hostile to life.

Acts of destruction have morphed into the destructiveness of normalcy. Having crossed sustainable yield thresholds of virtually all resources, humankind is destabilizing the very state of affairs whose stability lets it thrive in the first place. And this changes things, really in any direction we look. Before the shift, environmental harms came about by doing something, like grabbing an axe, going to a forest, and killing a tree. Now harms come about by doing nothing—just being alive is already sufficient.[3] This is paradoxical: perpetration of harm needs a perpetrator, and a perpetrator, to qualify as such, usually needs to do something. Not anymore: after the shift, sheer idleness suffices for perpetration.

The evaluation of actions has accordingly become puzzling. Presence alone creates harms. Furthermore, activity that would traditionally be regarded as virtuous, such as being a good citizen, raising a family, doing a job, and living an industrious and temperate life, can now have vicious effects: well-adjusted lifestyles, which used to be regarded as faultless, contribute to global warming. From a biospherical vantage point, the bum, the hippie, the artist, who loiters and loafs, who takes lovers instead of making kids, who smokes pot and lives a relaxed, bohemian life— a target of critique for Confucian philosophers and Protestant moralists alike—turns out to be the better person.

This counterintuitive outcome can even be quantified. A study comparing lifestyles, in leisurely Europe and industrious America, concluded that, "If, by 2050, the world works as many hours as do

Americans it could consume 15-30% more energy than it would following Europe. The additional carbon emissions could result in 1 to 2 degrees Celsius in extra global warming."[4]

New is also that effects smear out over space and time. This new feature of causation is a problem, because conventional normative theories are ill-equipped to deal with it. As D. Jamieson and J. Garvey have pointed out, "our usual paradigm collapses under the weight of certain complexities. Our values grew up in a low-tech, disconnected world of plenty. Now, cumulative and apparently innocent acts can have consequences undreamt of by our forebears."[5] Spatial smearing stretches causation over the entire planet. The US, for instance, constitutes barely the twenty-fifth part of global population yet causes a fifth of global GHG emissions per year, which spread in the atmosphere, with global effects.

Temporal smearing stretches causation over generations. The huge inertia and giant momentum of the fluids sloshing in the Earth System create a time lag. The warming felt in the first decade of the new millennium with the 2003 European heat wave, 2005 American hurricane season, or 2007 summer melt of 60% of Arctic sea ice, is the outcome of GHG emissions during the Kennedy administration.[6] Normal acts by normal adults thus steal their children's future—but they'll see it only when grown up.

Before the paradigm shift, it was as if we had given the Earth System a bad case of acne, despoiling its planetary face. With the shift, it is as if we're giving it a fever, provoking a geophysiological response aimed a purging the stricken system from its troubles. It is unclear to what extent feedback loops, regulatory mechanisms, and complex dynamic oscillations are like those in a living organism. But we are performing a global experiment to find out. The Earth, finding us intolerable, may shake us off.

Thus everything has changed. Environmental issues have grown into a causal avalanche, culminating in a single towering issue: the life-and-death risk of systems destabilization visible in climate change. The old conflict of nature and business was the tension between preserving beauty and making profit. Now profit is not at odds with beauty, but with survival—unless, from now on, profit-making will be limited.

Sustainability is the strategy of ensuring survival by limiting profit to returns within systems constraints. The term 'sustainability' has something to do with 'ability' to 'sustain'. It is a capacity to last or endure.

The German for sustainable strategies is *nachhaltig zukunftsorientierte Entwicklung*: a development (*Entwicklung*) that's future-oriented (*zukunftsorientiert*) such that it can keep going (*nachhaltig*). Sustainability is the 'keep-going' style of a process, or *Nachhaltigkeit*. This keep-going meme is old. It has many philosophical fathers. But Kant plays a special role. His ideas allow us to see sustainability in metaphysical depth. Kant reveals the geometro-dynamical pattern of the sustainable strategy, the humanistic context of this pattern, and the evolutionary direction this strategy is ultimately about.

§ 2

The Categorical Pyramid

Kant was not only a profound thinker, but also a visionary metaphysician with a lucky touch. His speculations sounded strange in his day, and their publication greatly delayed his university career, but they largely turned out to be right on the mark. He understood something about the dynamic structure of nature. His philosophy reflects this in the choice of topics. He was attracted to everything that snaps, crackles, and pops—and wondered what would happen to such dynamic events, their reciprocal dance, and their dialectical progression. He wrote on the interplay of energy and momentum (1747), terrestrial evolution (1754), and the moon brake of the Earth's rotation (1754). He wrote on the big bang, its cyclic nature, and the unfolding of life (1755). He wrote on oscillating power-points (what we would call superstrings) as units of matter and space

(1756). He wrote on earthquakes (1756) and storms (1756-57). He wrote on meteorites streaking across the sky (1764), and on volcanoes (1785). His ideas about evolution and emergence point to science in our times. His interest in all things energetic made him think about meteorological puzzles. He was the first modern philosopher who studied climate. Largely by sharp conjecture and creative imagination, aided by rational principles, he figured out how the monsoon works, what makes trade winds blow, and why coastal winds change direction.

Nietzsche had little patience with Kant's later critical writings but did admire his philosophy of nature. Likening Kant to a Presocratic, Anaxagoras—for Nietzsche the utmost in praise—he marveled about Kant's reduction of the being of nature to a 'moving mathematical figure,' or 'some kind of vibration'.[7] In retrospect, this identification of nature's essence has heuristic power. It explains much of the lucky touch of Kant's conjectures, and it remains a useful way of conceiving entities such as the Earth System. This 'moving mathematical figure' serves well as a pattern of mutually informing dynamic interplays, and it is the key to understanding Kant's approach to sustainability.

This approach is not like any other. It is philosophically distinguished, but it may also have been of personal significance. Kant was a weak and diminutive man, mired in poverty, ailing from a frail constitution, and who had no spouse to take care of him. He worked hard. Yet he lived up to eighty, nearly twice as long as the average European male of his century. And he lived well. Blessed with an enormous productivity, he turned out philosophical works for nearly sixty years; he started on his first book in 1744 and stopped working on his last in 1802. He even died well. Instead of falling into discomfort and disease, he dwindled until it was time to go. His exitus was as fine as could be.

Perhaps he was just fortunate. But then again, he clearly knew how to husband his physiological resources in a rational and efficient way. His exacting routine amused the town folk in Königsberg; after he turned middle age, it is said one could set one's watch by his daily afternoon constitutional. He organized his life like a clock work. And this clockwork just kept on ticking, for nearly eighty years.

The metaphysics of sustainability concerns clockworks that just keep on ticking. Their structure is articulated in Kant's Categorical Imperative. The Categorical Imperative is a theory of moral essence. This essence suggests a figure in spacetime. In time, the figure moves onwards from a point of origin. In space, this figure has three sides and a base. It looks like a three-sided pyramid, what mathematicians call a tetrahedron. All its faces are triangles. One forms the base. The other three are walls converging to an apex. From any point on the ground, one sees two sides at best. From any point in space, one sees at most three. Seeing all of them needs one to move around the pyramid. The moral essence, encapsulated in the Categorical Imperative, involves three formulas resting on a foundation.

The foundation is sound action. A sound action is informed by practical reason and good will. Practical reason is the geometry of action; good will is the dynamics of action. The moral pyramid rises from a geometro-dynamics of actions based in practical reason and good will. On this base are three sides that come together in a point. The three sides are sustainability, humanity, and evolution. The point is the good. Sustainability is the front of the pyramid; it's the face one sees first. For Kant, this is its formula:

> So act as if the maxim of your action were to evolve by your will to a universal natural law.[8]

An action evolves over time. It has a beginning and an end. The beginning is an idea, an intention, or a decision to do something—a plan of action or a maxim. The end varies; it depends on whether the maxim works out or not. When the maxim works out, the action evolves to a universal natural law. This is the best possible outcome. Actions that lead to such outcome are, by definition, moral. The end of moral action is the realization of the maxim in optimal form. In space, the action is done everywhere.

In time, the action is done steadily. In this way, the action harmonizes with the rhythm of life. This is what evolution to a universal natural law means. Between the beginning and the end of this evolution is a path of progressive realization of the maxim. The enactment is a development marked by stages. The first stage is the individual enactment. One person executes a plan and performs the action. The next stage is a collective enactment. Other persons follow the example of the first person and copy the same behavior. The final stage depends on whether this works or not. If the collective enactment works out, and the action, in retrospect, reveals itself as doable, collectively, in space and over time, the final stage is the endless repeat of the second stage. The end of evolution is an ongoing steady state. People keep doing what the first person had done; the action has become 'the thing to do'. One could think of the maxim like a seed germinating into action, and of actions getting copied until they spread in the scenery. That's sustainability. Its formula is the imperative: act so that your act can be copied.

Sustainable actions are moral actions by definition. From an environmental point of view, we would say that sustainable actions are the right thing to do. Kant agrees with the environmental claim but also turns it around and thus goes a step further. In his point of view, the right thing is always sustainable. What we call sustainability Kant called universalizability (the capacity of a plan of action to be 'universalized'; i.e., to replicate consistently into the thing to do), which is synonymous with the good. Sustainability blends with the good just as unsustainable actions carry the mark of evil. Formally, this is how one recognizes evil: when acted out collectively, such ideas cannot be copied continuously. Copying evil generates conditions that prevent more copying. Evil, when replicated, breeds issues that first slow down and eventually terminate enactment. The copies can spread temporarily, but the spread is unstable. They're like infectious diseases running rampant, which eventually wind down. Or like fires that burn themselves out. The mark of evil is that its maxim fails to evolve to a universal law.

An example would be conventional US lifestyle, the least sustainable way of life on the planet today, and the chief driver of global warming. The American four percent of world population is responsible for twenty percent of annual global GHG emissions. In cumulative terms, the US is guilty of more than a third of anthropogenic GHG emissions. Suppose your maxim is to act like a typical US Republican. Could you rationally will this maxim to evolve to a universal natural law? You cannot. Just do the math. If the other 96 percent of humanity emitted as much GHGs as Americans do at present, then global emissions would increase by a factor of 4.8. If this had happened cumulatively, constituting full climate forcing over time, then atmospheric GHG concentrations would now be 4.8 times higher as they are—instead of 460 parts per million in a volume of gas (ppmv) of the total CO_2-equivalent concentration of long-lived GHGs, we would measure 2,208 ppmv CO_{2eq}.[9] At that GHG concentration, the Earth System would sharply fall out of balance and climate change would accelerate to runaway cascades. A spike of temperatures similar to the one in the Permian-Triassic extinction event would be likely. Civilization would collapse. Humankind would die back. If everyone acted like Americans, then, eventually but inevitably, *no one* would be able to act as Americans do. Evil tends to be self-reducing.[10]

§ 3

Sustainability, Evolution, and Humanity

The three sides of the moral pyramid, as mentioned, are sustainability, humanity, and evolution. Humanity, the second face, supplies the context to sustainability. Evolution, the third face, lends it direction. Context and direction matter. They complete the metaphysics of sustainability. The sustainability-formula, as seen, is a 'keep-going' figure. This matrix of universalizing acts into patterns is grounded in existential realities. The reality of Dasein is us. Kant puts it this way:

> Act so that you treat humankind, as much in your person as in the person of every other, at all times simultaneously as an end and never as a means only.[11]

It is natural to use one another and ourselves as means to certain ends. We do it all the time, and we have to. We need each other. So we use each other. This is fine as long as such use is coupled with respect; as long as we treat ourselves and one another not only as means but also as ends. But what's not fine is to treat humans as means only. This is the second dimension of the Categorical Imperative. Let's call it the humanity-formula.

Interesting about this humanity-formula is that it is not a superficial moral exhortation. The idea is not to respect humanity for its own sake. Although the Categorical Imperative is occasionally read as the suggestion of some sort of dogmatic anthropocentrism, by environmental philosophers in particular, there is no need to stay with such an interpretation. For one thing, the humanity-formula has a reason. For another, this reason is not ethical, but ontological instead. We ought to treat humans as ends because they really are ends. That is to say, humans are self-directed, goal-oriented, and highly self-determined living beings, compared to other species on the planet. In comparison, humans are the best embodiment of autonomy. No other species enjoys the degree of freedom of action we do. In this sense, humans are ends. Ants are means to the survival of the hive. Birds and nonhuman mammals are means in that they're largely at the mercy of their instincts, and because their freedom seems limited.

There is also nothing personal or 'speciesist,' as P. Singer would say, about treating humanity as an end. It just so happens that in our biosphere the human species is the closest biological approximation to the idea of autonomy. The humanity-formula is not dogmatically about humanity. It applies to any living entities that are autonomous. The point then is to treat autonomous beings as autonomous beings. The way to do so is to accord them the respect appropriate to beings able to make up their own minds. It is inappropriate to treat them as means only. Doing so would betray a lack of courtesy. Doing so would also indicate blindness to the nature of autonomy, a lack of recognition of a given constitution.

This courtesy should also be extended to members of the set of autonomous beings who are, in some form or other, impaired. The sick and handicapped in need of care givers are not autonomous, but they could be autonomous, and would be if they recovered. Children in need of parental supervision are not autonomous but will be when they grow up. The elderly in need of assistance are not autonomous but had been when they were younger. Potentials matter in geometro-dynamics. And they matter to courtesy. It is inappropriate to treat elders as if they'd never been younger, or children as if they'd never grow up, or the sick and handicapped as if recovery is inconceivable. It is inappropriate also on the level of cognition, because the stated 'as-if' scenarios happen to be unrealistic. Thus the humanity-formula applies to all humans without this being shallow anthropocentrism.

The appropriateness of courteous treatment points to factual states-of-affairs. There is an ontological difference between entities that are actually or potentially autonomous and those that are not. Using humans as mere means is a cognitive mistake that, when enacted, introduces an inefficiency that can spread through a matrix of actions. Compensating inefficiencies needs energy. Hence enslaving people takes force, and maintaining slavery requires continuous exertion of power. Efficient actions are those whose performance is done cognizant of the context in which the performance is to be located.

Humanity is the context of sustainability. Unsustainable actions, as illustrated by conventional US lifestyle, involve the exploitation of resources such as carbon sinks beyond their capacity for renewal. This degrades the resource. The generation of humans that exists today uses resources unsustainably and thus takes them away from the next generation. Such a theft from the future is an abuse of descendants. Future offspring are treated as means only—what's theirs is ours, and what's ours is none of their business. The context of humanity reaches in space across the globe, and stretches in time over generations. Disregarding this intergenerational context breaks the ancestral chain. We are like greedy elders burning through our wealth, leaving nothing for the children. When elders scorn their children, the children, when grown up, have reason to spurn the elders who betrayed them before passing on. This betrayal breaks the bond. Our unsustainable conduct at present is likely to prevent us from ever becoming ancestors worthy of respect in the eyes of the next generation. Odds are they'll despise us because we fail to look ahead.

The temporal context of sustainability thus acquires direction. This direction is an orientation to the future. The way of sustainability, in the context of time, is a forward-looking progression. Humanity is to be treated as an end, and this means everyone in space *and* in time. An action is moral when it can be replicated indefinitely across space and through time. This self-generating progression is the third and final face of the moral pyramid. Kant puts it this way:

> Act in harmony with intentions of any legislative member for a possible set of ends (4:439.1-3).

Legislative members are autonomous individuals. A possible set of ends is a realm of autonomy. The intentions of any legislative member for a possible set of ends are to enhance, enrich, and add on to the set. Legislative members of the end-set doing the opposite would act self-contradictorily. Thus autonomy begets autonomy, and complexity begets complexity. The right thing to do is to treat reality like a garden of information, and to work as a rational gardener, who makes information flourish. Evolution, in the post-Darwinian sense of emergence, is accordingly the final thrust of sustainability.

While this sounds visionary—perhaps unavoidably so—the sensible substance of this formula will come into view if one first considers doing the opposite and next looks at mundane reality. The opposite would be to act contrary to intentions of legislative members for a possible set of ends, hence to do something that fails to evolve to a universal natural law. It would be inherently self-reducing. No legislative member can rationally will this. Doing so is stupid, and, as we've seen, it is evil.

It is also unnatural. It would go against the grain of things. The sensible substance of the evolution-formula comes into view when looking at mundane reality. The universe has existed for 13.7 billion years. In this time, simplicity has unfolded into increasing complexity. Ever evolving limits of nature permit the progressive production of the matters on the periodic table, from the plainest, hydrogen, to the most elaborate and fleeting, the transuranic elements. The Earth has existed for 4.6 billion years, and life on Earth is 3.5 billion years old. The ever evolving constitution of the Earth System allows for the birth of organisms with ever longer genetic chains. Our ancestors were simple. As time goes by, biotic information grows more refined and more complex. In the Standard Model this is nature's way.

The evolution-formula lends direction to the imperative of sustainability in its humanity-context. This direction is moral, of course, but it is also profoundly *natural*. It is indeed the way of events happening. Thus toiling as a rational gardener in the plot of information just means to accord with nature's way, and to do what comes naturally indeed. Information wants to flourish. Ergo we should help it along.

A practical aspect of this direction is mirrored in policy. When dealing with environmental dangers, efficient policy is evolutionary policy. When a catastrophe happens that requires repair and rebuilding, rational policy dictates the construction of structures that are more resilient and safer, instead of simply rebuilding the structures that had been destroyed. Failing to eliminate known vulnerabilities in the affected structures is an accident waiting to happen, since it merely invites the next disaster to occur. When an earthquake levels a town, buildings are consequently seismically retrofitted when rebuilt. When a hurricane floods a city, levees and dikes are rebuilt beyond their original size and height. Sustainable actions are keep-going patterns that persist by their evolving attunement to realities.

§ 4

Hardcore Theodicy

> *That's a question I can't answer. For the hundredth time I repeat, there are numbers of questions, but I've only taken the children, because in their case what I mean is so unanswerably clear. Listen! If all must suffer to pay for the eternal harmony, what have children to do with it, tell me, please? It's beyond all comprehension why they should suffer, and why they should pay for the harmony. Why should they, too, furnish material to enrich the soil for the harmony of the future? I understand solidarity in sin among men. I understand solidarity in retribution, too; but there can be no such solidarity with children.*
>
> *And if it is really true that they must share responsibility for all their fathers' crimes, such a truth is not of this world and is beyond my comprehension.*
>
> Ivan, to Alyosha
> *Brothers Karamazov*, ch. 4
> Tr. Constance Garnett

Keep-going patterns and their evolving future attunement open larger vistas about good and evil. Contemplating the rise and fall of cultures makes one wonder about the justice of it all. How to rate the unsustainability of the present age, however, is a dull question to ask, since the answer would be as obvious as it would be biased. Of course we'd deplore this, and think with a heavy heart of the fate we make for children, since they're human, and we're human, and it's natural to feel solidarity with kin.

How the verdict would fall out in the absence of solidarity is a question to pursue. How would a distant observer rate the events? Consider a culture rising from our eventual ashes, a culture of nonhumans. Whether they'd be descended from birds, dolphins, or bees won't matter. Whatever their genome might be, let's imagine their perspectives would be intelligible to us, because these beings, too, would be social yet autonomous, child-bearing, and of varying levels of sensibility and intelligence. Let us assume that theirs is an open society, in which these beings exchange information and debate opinions, allowing majority views to emerge without constraints. What would their likely consensus on us be?

Such beings, peering into the past, would see the unsustainable conduct of our species at present. They would see us approaching a fork in the road. One path would steadily rise upwards, progressing to a sustainable world. Another would spiral downward, regressing to wars and ending in extinction. If they saw us taking the former, they'd likely applaud this rational resolve and civil evolution. How could they not? All species face the same challenges, and it beautiful to see that some rise to the occasion.

On the other hand, how would we seem, in their eyes, if we failed at the attempt at a civil evolution? Suppose we regressed to the scenario feared most by the IPCC, with the sharpest rise in GHG emissions and world temperatures, a world of Republicans and Taliban, a "very heterogeneous world," whose "underlying theme is self-reliance and preservation of local identities"; a world in which "fertility patterns across regions converge very slowly," resulting in "continuously increasing population," and where "economic development is more fragmented and slower than [in] other storylines."[12] This vista (scenario A2 in IPCC parlance) leads to a self-reducing, self-terminating society. We would fade into the night. Our legacy would disappear with us.

Would they morn us? Odds are they wouldn't. We had a chance, they might say, and we didn't take it. Well, then, we had it coming. Tragic are the inexorable workings of fate that doom a hero regardless of his struggle. But there's nothing heart-rending about going down because one acts in ways that cannot be universalized, that disregard humanity, and that go against the grain of things. Who morns Hitler's ephemeral empire of inhumanity? So why should anyone morn yet another fever blossom driven up and brought down by unsustainable policies? This is not tragic. It is merely stupid.

This, then, would be the fork and the verdict by distant observers. Either we make it and embrace sustainability, or we don't. And either way, it's all good. Praiseworthy are those whose actions are universalizable. Others, who prepare their own demise despite knowing better, are no great loss.

And yet, even if this became the majority opinion of an inhuman civilization assessing our fate, one question would likely reverberate through this culture, ringing with sadness: what about the children?[13]

Endnotes

1. Gro Harlem Brundtland, World Commission on Environment and Development (Brundtland Commission), *Our Common Future* (Oxford/New York: Oxford University Press, 1987), a report welcomed by the UN in the General Assembly resolution 42/187 (December 1987).
2. E. Assadourian, ed., *2010 State of the World: Transforming Cultures from Consumerism to Sustainability*. A Worldwatch Institute Report on Progress toward a Sustainable Society (New York/London: Norton, 2010); cf. also, L. Starke, ed., *2009 State of the World: Into a Warming World* (New York/London: Norton 2009); D. Esty, ed., *2008 State of the World: toward a Sustainable Economy* (New York/London:

Norton, 2008); L. Starke, ed., *2007 State of the World: our Urban Future* (New York/London: Norton, 2007); L. Starke, ed., *2006 State of the World Report: Special Focus China and India* (New York/London: Norton, 2006).

3. James Lovelock, in *The Vanishing Face of Gaia: a Final Warning* (New York: Basic Books, 2009), 5, put this point nicely: "No voluntary human act can reduce our numbers fast enough even to slow climate change. Merely by existing, people and their dependent animals are responsible for more than ten times the greenhouse gas emissions of all the airline travel in the world."

4. D. Rosnick, M. Weisbrot, *Are Shorter Work Hours Good for the Environment? A Comparison of U.S. and European Energy Consumption* (Washington, DC: Center for Economic Policy and Research, 2006), 1.

5. James Garvey, *The Ethics of Climate Change: Right and Wrong in a Warming World* (London: Continuum, 2008) 58-19; cf. also Dale Jamieson, "Ethics, public policy, and global warming," in Jamieson, *Morality's Progress: Essays on Humans, Other Animals, and the Rest of Nature* (Oxford: Clarendon Press, 2002), 291-293.

6. S. Faris, *Forecast: the Consequences of Climate Change, from the Amazon to the Arctic, from Darfur to Napa Valley* (New York: Henry Holt, 2009), 222.

7. Nietzsche, *Die Philosophie im tragischen Zeitalter der Griechen* # 17, Nachgelassene Schriften, in: *Sämtliche Werke: Kritische Studienausgabe*, ed. G. Colli, M. Montinari (Munich: DTV, 1980), 15 vol., 1:866-867.

8. Kant, *Sämtliche Schriften* (Academy Edition), 4:421.18-20

9. The atmospheric CO_2 concentrations for 2009 were 382 ppmv. GHG, of which CO_2 is one, are measured in CO_2 equivalents. This unit of measurement allows one to compare climate effects of all GHGs against one another. CO_{2eq} are calculated by multiplying the quantity of a greenhouse gas by its global warming potential. 460 ppmv CO_{2eq} are the IPCC estimate for 2007. Cf. W. L. Hare, "A Safe Landing for Climate," box 2-2 "Greenhouse Gas Concentrations and Global Warming," in Starke (2009), 23, and "Glossary," ibid. 201.

10. A poetic rendition of Kant's idea is in the Bible, Romans 6:23, "the wages of sin is death". If 'sin' is understood as evil actions, and 'wages' as their consequences, then 'death,' so interpreted, would be the self-termination of such actions. The problem with the Bible verse, however, is that isolated instances of actions unsustainable in principle (and thus evil by definition) are not self-reducing. The pattern only emerges in steady collective instantiation.
11. Kant, 4:429.10-12.
12. S. Solomon, D. Qin et al., "Summary for Policy Makers," in *Climate Change 2007: the Physical Science Basis*, contribution of Working Group I to the Fourth Assessment Report of the Intergovernmental Panel on Climate Change (Cambridge, UK: Cambridge University Press, 2007), p. 18; see the A2 scenario in box "The Emissions Scenarios of the IPCC". By contrast, lowest emissions would happen in the (best-case) B1 scenario, which "describes a convergent world with the same global population, that peaks in mid-century and declines thereafter ... with rapid change in economic structures toward a service and information economy, with reductions in material intensity and the introduction of clean and resource-efficient technologies. The emphasis is on global solutions to economic, social and environmental sustainability, including improved equity, but without additional climate initiatives."
13. The author wishes to thank the participants in the 2010 climate philosophy seminar for inspiration to this essay.

CHAPTER TWO

Sustainability: A Personal Account[*]

J. Baird Callicott

(1) Why is sustainability a contested concept? (I did not surround "sustainability" with quotation marks because, as the first question indicates, the concept not the term is what is contested.)

In my opinion, sustainability is fraught with associative conceptual baggage, which has created a penumbra of suspicion around it.

Classic natural resource management was focused on single species—this kind of game animal, that kind of commercially harvested fish—the goal of management being to attain *maximum sustained yield* of the target "resource." The quantitative formulae for calculating maximum sustained yield were appropriately sophisticated mathematically—involving constants for the species' fertility, growth rate, age of reproductive maturity, and so on. They were not, however, at all sophisticated ecologically. Little attention was paid to the feedback loops between those species that strongly interact with the species of interest and even less attention was paid to those other species that strongly interact with species that strongly interact with the species of interest. As a result, the idealized steady-state or predictably oscillating logistic equations of population growth, stabilization, harvest, and rebound bore little resemblance to the behavior of actual populations of natural-resource species on the ground or in the oceans. The mismatch between classic population model and actual population fact is attributable to the impact of artificially fluctuating

---

[*] Reprinted by permission of the author (J. Baird Callicott) and the publisher (Automatic Press/VIP). From *Sustainability Ethics: 5 Questions*, eds. Ryne Raffaelle, Wade Robison, & Evan Selinger (Copenhagen: Automatic Press/VIP, 2010): 57-70.

populations of the target species on other species in the biotic communities of which they are members and, in turn, the unpredicted (and perhaps unpredictable) impact of their fluctuating populations on the species of interest. So, the concept of sustainability is tainted by association with the older "Resourcist" concept of maximum sustained yield. Or even forget "maximum"; it is tainted by Resourcism itself, the concept of sustained *yield*—as if the natural world is nothing but a "pool" resources existing to service human wants as well as needs.

In response to the politically sensitive tension between economic development for impoverished human populations and the destruction of natural capital that, historically, accompanies human economic development, the concept of sustainable development was offered as a resolution. Perhaps it would be possible to eat our environmental cake and have it too—by means of *sustainable* development. Many skeptical environmentalists of a Malthusian bent regard "sustainable development" to be an oxymoron. If poverty is defined as a low standard of living and a low standard of living is equated with low levels of consumption of goods and services, then economic development for impoverished human populations entails increased consumption of goods and services, which in turn entails increased conversion (destruction) of natural capital on a finite, small planet—or so the neo-Malthusians grumble.

Personally, I am not convinced that either the concept of sustained yield or that of sustainable development is fatally flawed. Calculating the former is a technical problem—a daunting one to be sure, but a technical problem nevertheless. Increased mathematical sophistication by, for example, the use of nonlinear equations; increased modeling sophistication by, for example, the use of multi-agent-based models; and the ever increasing computational power of evolving computer hardware and software may make it possible for modelers to approximate the behavior of species populations of interest in a complex environment of coupled natural and human systems. And taking the concept of sustainable development seriously might require challenging the rather conventional assumptions that its skeptics make. Might an impoverished people's standard of living be increased without a corresponding and proportionate increase in its levels of consumption of goods and services? Possibly, I would think. Can increases in the consumption of goods and services be attained without a corresponding and proportionate increase in the conversion of natural capital? Again, possibly, I would think. And here too, the problems are partly technical. But they are also partly psychological, cultural, and political. Better agricultural techniques, for example, might increase food production without converting more land to crops; alternative techniques of generating energy might raise standards of living without adversely affecting the environment. Re-envisioning the good life—from consumerist values to those of association, environmental aesthetics, education, vegetarianism—might increase standards of living without an inevitable adverse impact on the natural environment.

But these more optimistic ruminations of mine are neither novel nor original. To get back to the question at hand—why is sustainability a contested concept?—I think, in short, that one part of a complete answer to that question is this: The concept of sustainability is presumed guilty by association with the Resourcist concept of (maximum) sustained yield and the putatively pollyannaish and allegedly oxymoronic concept of sustainable development.

(2) How is your preferred definition of "sustainability" better than alternative accounts? (Here I add the quotation marks because terms, I take it, have definitions, while concepts have characterizations.)

The answer to this question requires posing two others: (a) what is my preferred definition of "sustainability?"; and (b) what are the alternative accounts? Actually, I would reshuffle these subordinate questions in the following way: (a′) what are alternative *definitions* of "sustainability"?; and (b′) what is my alternative *account* of the concept. After answering these sub-questions, I'll turn to a reshuffled main question (2′): How is my preferred *account* of sustainability better than alternative *definitions* of "sustainability?"

Sustainability is a property of an activity or complex system of activities capable of going on and on indefinitely, if not forever. As my answer to the previous question indicates, sustainability is usually associated with human activities or complex systems of human activities that are constrained by *environmental* limits. But any human activity may or may not be sustainable, many involving no proximate environmental constraints at all. Gambling losses of a thousand dollars a week, week after week, are not sustainable by a person who earns fifty thousand dollars a year. A national health-care system the cost of which increases by ten percent per year is not sustainable.

Because the property of sustainability is a property of activities—processes—the concept of sustainability implicitly, but obviously, references time. Further, although the temporal scale of sustainability is rarely specified, one unconsciously and often very vaguely scales the temporal parameters of sustainability relative to the human activity or complex system of human activities in question. Continuous running is a human activity that may or may not be sustainable, but the temporal scale of sustained continuous running is calibrated in minutes and hours. I would suppose that a national health-care system that is designed to function well for a century would be deemed sustainable. But I would also suppose that an agricultural system would be deemed unsustainable if it was designed to function well for only a century, after which period it would collapse. One would probably not, however, withhold the sustainability descriptor from an agricultural system that functioned well for a million years followed by collapse. Where should the temporal parameter of sustainability be drawn for an agricultural system? At a thousand years, two thousand, ten thousand . . . ? I don't know. My point simply is that the temporal scale of sustainability is not infinity and, though often both implicit and vague—as the word "indefinitely" in the first sentence of the previous paragraph is meant to suggest—it is relative to the human activity or complex system of human activities in question.

There is, of course, a very famous definition of "sustainability," or, more precisely, of "sustainable development"—that of the Brundtland Report: "development that meets the needs of the present without compromising the ability of future generations to meet their own needs." What is striking about this definition is that it makes no reference whatever to environmental quality. Consider an alternative: development that meets the needs of the present and those of future generations without compromising environmental quality—without polluting the atmosphere, oceans, or freshwaters; without causing mass species extinction; without laying waste to Earth's rainforests; . . . . The Brundtland Report definition of "sustainable development" is laudable because it implicitly distinguishes between *needs* and *wants*, but, as things presently stand in our consumerist culture, satisfying wants is thought to be a need. So, even the Brundtland Report's laudable focus on needs is less than satisfactory, in my opinion, without explicitly divorcing needs from wants.

Doesn't the Brundtland Report definition at least imply sustainably harvesting renewable natural resources (sustainable yield) and preserving ecological services, without which future generations could not meet their needs? It may not mention environmental conservation explicitly, but it implies the necessity of sustaining environmental quality as a condition for future generations meeting their own needs, doesn't it? No. It doesn't. In neo-classical economics there is the substitutability axiom. As a heavily exploited natural resource becomes scarce its price increases, making investment in finding a substitute become increasingly attractive. Thus, there is no need to conserve any particular natural resource. For example, when we begin to run short of copper for making telephone wires, someone will (as someone did) invent fiber optics. Such accumulated anecdotal evidence suggests that market forces will always stimulate the discovery or invention of substitutes for any natural resource— from petroleum to Madagascar periwinkles. As one species of marine finfish is harvested to commercial extinction the fishing industry just exploits another; when all marine finfish stocks are depleted, someone will figure out how to make sushi with a fish-flavored and fish-textured soy product. According to this prevailing way of thinking, the present generation can, therefore, meet its own needs—including the presently prevailing need to satisfy wants—by rapidly exploiting existing organic natural resources to commercial if not to biological extinction and bequeathing a legacy of wealth and technology and a culture of business and inventiveness to future generations—by means of which they can meet their own needs.

Another, more recent definition of "sustainability" is offered by Bryan Norton in his book, *Sustainability*. Norton rejects what he calls "weak sustainability," the economistic understanding of sustainability that the famous Brundtland Report definition invites. He also rejects what he calls "strong sustainability"—the view that artificial capital cannot substitute for natural capital—championed by ecological economists and conservation biologists. What they regard as a foundational natural legacy for future generations—biodiversity, for example—Norton demeaningly refers to as "stuff." Advocates of strong sustainability think it necessary to bequeath some natural "stuff" to future generations. Norton goes on to define "sustainability" in terms of a less tangible legacy: "*sustainable* and *sustainable development* are not themselves general *descriptors* of states of societies or cultures but rather refer to many specific sets of commitments on the part of specific societies, communities, and cultures to perpetuate place-based values and project them into the future."

What those place-based values are is up to specific societies, communities, and cultures to determine democratically. While I certainly agree with Winston Churchill that "democracy is the worst form of government, except for all the others," I am less sanguine than is Norton that place-based values will always project environmental quality into the future. That may happen in some specific societies, communities, and cultures—say Ashland, Oregon and Boulder, Colorado. But it may not happen in others—say Talladega, Alabama and Odessa, Texas, where the place-based values likely to be projected into the future center, respectively, on NASCAR racing and wildcat drilling for the petroleum industry's gold standard—light, sweet Texas crude.

Further, Norton's implicit bioregionalism, a geographical ontology of "place" and a sociological ontology of *"specific societies, communities, and cultures"* is quaint. The twenty-first century is characterized by globality—for better or worse. Our most daunting and urgent environmental challenge is *global* climate change. Specific places, societies, communities, and cultures can no longer be considered in isolation from one another and can no longer plausibly make independent commitments to sustainability. Without international cooperation to mitigate carbon emissions, no place will look like the same place fifty years from now, no matter what commitments to place-based values its denizens make and project into the future.

Twenty-first century ecologists have also abandoned the concept of closed, self-regulating, ontologically robust ecosystems. Ecosystems are partly artifacts of ecological hypotheses. Once identified as such, they are porously open to all sorts of comings and goings, from invasive organisms to minerals from afar blown in on the wind and washed down by the rain. For example, the scant fertility of Amazonian soils depends on African dust blown across the Atlantic on the prevailing easterlies. If Africans managed to reverse desertification and stabilize their soils, in a successful effort to achieve regional agricultural sustainability, the Amazon rainforest would be starved for nutrients, with a huge loss of biodiversity. Sustainability has an implicit spatial scale, as well as an implicit temporal scale, and it is becoming increasingly clear that that scale is global, not local. Sustainability is thus not a matter of independent commitments on the part of "specific societies, communities, and cultures" as per Norton's definition.

I would characterize sustainability as follows. The human economy is a subset of the economy of nature. Need it be recalled that the words "ecology" and "economy" were coined from the same Greek root, *oikos*, meaning "home"? Sustainable human economic activity would not disrupt the globally integrated ecological processes and functions of our global home, planet Earth. Obversely put, sustainable development consists in devising artificial ecosystems (human economies) that are modeled on and symbiotically adapted to the economy of nature (the global ecosystem, the living biosphere). The macro-economy of nature is the model for sustainable human economic microcosms.

A good, explicit example of devising artificial ecosystems that are adapted to and modeled on natural ecosystems is the perennial polyculture envisioned by Wes Jackson and in process of development at the Land Institute in Salina, Kansas. Jackson's model is the prairie, which, under various names (grassland, savannah, steppe) on various continents (North and South America, Asia, Africa) is a globally distributed biome. According to Jackson, four kinds of plants must constitute a perennial polyculture: cool-season grasses, warm-season grasses, legumes, and sunflowers—all perennialized. If successful (or rather when successful, as progress has been notable), Jackson's would (will) represent a second agricultural revolution, the first, based on annual monocultural grasses, having occurred worldwide about ten or eleven thousand years ago. Another example is the suite of artificial ecosystems going under the rubric of "industrial ecology," the general idea being that the waste product of one industry is the resource of another. Spent fry oil, to take but one instance, from fast-food restaurants can be (and is in some places) the feed stock for the manufacture of biodiesel fuel for the diesel engines of cars, trucks, and buses.

My preferred characterization of the concept of sustainability is better than the famous definition of "sustainable development" in the Brundtland Report and the definition of "sustainable" in Norton's prominent tome for the reasons already stated.

(3) What is sustainability ethics, and how does it differ from more established forms of applied ethics, such as environmental ethics and business ethics?

I would say that sustainability ethics is the interface between environmental ethics and business ethics. My preferred understanding of sustainability revolves around the concept of economy—the planetary economy of nature and the globalized human economy. Sustainability, as I prefer to think of it, is a matter of adapting human economic systems to and modeling them on the economy of nature in which the globalized human economy is embedded and in relation to which it should stand as microcosm to macrocosm. This way of understanding sustainability comes out of environmental ethics. At least, so I must suppose, because I am an environmental ethicist and this way of understanding sustainability is a product of my reflections and cogitations as an environmental ethicist. To actually achieve sustainability, so understood, would seem to me to be a matter not of environmental ethics, but of business ethics.

If the concept of economy is what my understanding of sustainability revolves around, why would actually achieving sustainability not be a matter of economics, instead of being a matter of business ethics? Because economics is a descriptive social science, with predictive ambitions based on a number of controversial assumptions, among them the invidious axiom of substitutability. Also among those controversial assumptions are that all values are preferences and that human welfare consists of maximizing "preference satisfaction." To treat all values as preferences enables economists to quantify all values in a monetary metric for purposes of comparison and thus to make economics a totalizing discipline, beyond the reach of which there is nothing. But all values are not preferences. I may prefer strawberry ice cream to chocolate, but I do not prefer human emancipation to slavery. Quite the contrary, I would actually prefer to have a few slaves at my beck and call. However, I and any other moral human being would refuse to own slaves, if given the opportunity to do so, because slavery is a gross violation of human dignity and human dignity is something moral human beings value, not something we merely prefer. Values proper constrain preferences and values proper should not be subjected to shadow pricing and benefit-cost analysis. Nor does human welfare consist of maximizing preference satisfaction; to think that it does is to think like a two-year old.

Business ethics consists precisely in exploring various values proper—as opposed to preferences—and recommending that some such values negatively constrain and positively inspire and guide human economic activities. I am not very familiar with what business ethics is all about, but I would imagine that it would, at a minimum, recommend that human economic activities be constrained by the aforementioned value of human dignity, which would, among other things, prohibit cost-cutting by employing child or prison labor in workplaces that are a hazard to life, lung, liver, or limb. I would also imagine that business ethics would inspire and guide human economic activities by recommending that businessmen and -women set a goal of enhancing the general welfare (not understood in terms of preference satisfaction), which decent people value, as well making a private profit.

At the interface of environmental and business ethics, environmental values proper would be recommended as constraints on and inspirational guides for human economic activity. Such values would range from a concern for animal welfare to clean air and water to biodiversity conservation. Sustainability ethics, as I conceive it, is business ethics informed not only by the extra-economic values of traditional social ethics, but also by those of contemporary environmental ethics, all in service of the goal of achieving sustainability—that is, atuning the human economy to the economy of nature.

(4) What unique contributions can the discipline of philosophy make toward enhancing our understanding of what sustainability is and how sustainable goals can be accomplished? (Here I substitute "toward" for "towards" because "towards" is, I believe, a Britishism.)

The foregoing is a contribution within the discipline of philosophy toward enhancing our understanding of what sustainability is, primarily, and, secondarily, toward how sustainable goals can be accomplished. So let's stand back and look at what is going on.

First, the discipline of philosophy is concerned with conceptual clarification and precision of expression. Sometimes philosophical conceptual clarification and precision of expression can be annoying because it may seem too fastidious, even trivial.

Note, for example, my possibly annoying distinction between concepts and the terms that we use to name them and how concepts are appropriately characterized in an account, while only terms are appropriately defined. These incidental issues of conceptual clarification and precision of expression—the clarification of the difference between concepts and terms and definitions and accounts and the determination to express them all precisely—may seem (and even be) trivial. But clarifying the difference between the concepts of sustained yield, sustainable development, and sustainability goes to the substantive heart of the matter.

Second, the discipline of philosophy is concerned with critical thinking. I subjected the definition of "sustainable development" in both the Brundtland Report and in Norton's book to extensive critical examination and clearly exposed the inadequacies of each. I subjected the assumptions of mainstream economics to a more cursory regimen of critical thought.

Third, the discipline of philosophy is concerned with values. In other disciplines, especially those with scientific pretenses, values are often treated as purely subjective, personal, arbitrary, and irrational. Preferences may be all these things, but, as noted, not all values are preferences. And values are of the utmost importance. Whether we admit it or not, values drive all human activity and all public policy. Recognizing values for what they are, exposing them to people who have been taught to marginalize and trivialize them, and stressing their ultimate importance is a central and unique contribution that the discipline of philosophy can make to any human endeavor, including enhancing our understanding of what sustainability is and how sustainability can be achieved.

Fourth, the discipline of philosophy is concerned with ontology, with what exists and what does not. I criticized Norton's ontological assumptions about geography and society, his hypostatization of place and specific communities. To make more explicit another ontological issue lurking in the foregoing discussion, let me ask: Do ecosystems exist robustly—as robustly as say a snake or a monkey? I hinted that they do not. They exist, but their existence is less robust than that of snake or a monkey. That ecosystems have fuzzy boundaries is the least ontologically problematic thing about them. More problematic ontologically, their (often fuzzy) boundaries are determined by the particular interests of the ecologists who investigate them. In the instance mentioned in the foregoing discussion, the Amazon rainforest is taken to be an ecosystem sustained by dust blown in from Africa. But if an ecologist were interested in energy flows in Amazonian food webs, the boundaries of the ecological object of study would be drawn very much more narrowly. Ecosystems are thus partly scientific artifacts, they come into being, partly, only when interrogated by ecologists.

Finally, the discipline of philosophy is concerned with epistemology—with what we know, how we know it, and the limits of human knowledge. Scientific knowledge is often privileged epistemologically in comparison with other knowledge claims. At the same time, uncertainty is becoming an ever more prominent epistemological issue in science itself, an epistemological issue that is also ever more prominent at the interface of science and policy and politics. Epistemological issues hardly appear at all in the foregoing discussion, but perhaps they should have taken more of a center-stage position. Who gets to declare what is and what is not a sustainable human activity? Perhaps unconscionably, I disparaged the epistemic credentials of Talladega rednecks and Odessa roughnecks. Perhaps arrogantly, I dismissed without hesitation, their various claims to know what is and what is not sustainable. A signal and laudable virtue of Norton's definition of "sustainable" is that it is not so dismissive as is my account of sustainability. But should every claim to knowledge be treated with respect and given an equal hearing with every other? What about the knowledge claims of Scientologists who believe that they arrived on a spaceship from another planet or evangelical Christian fundamentalists who claim to know that the Earth has only been around for six thousand years? Aren't some extra-scientific knowledge claims just as worthy of contempt as others are of respect? How do we determine which knowledge claims warrant appropriate respect and which warrant appropriate contempt? That is an epistemological problem; and epistemology is a subdiscipline of philosophy.

(5) What are the most important topics of future inquiry that sustainability theorists need to investigate?

Assuming that sustainability requires a multidisciplinary approach and that sustainability theorists inhabit a wide variety of disciplines, then the topics of future inquiry might be sorted by discipline. In the foregoing discussion I mentioned two disciplines in which topics of future inquiry are quite technical: perennial polyculture agriculture and biofuels engineering. In the case of other disciplines, economics, for example, sustainability challenges some very deep assumptions: substitutability, for example; the reduction of all values to preferences, for another which, economists assume, can all be quantified in a monetary metric; the understanding of what it means for human beings to fare well in terms of preference satisfaction. In addition, sustainability challenges the practice by economists of discounting future benefits at the current rate of interest. Sustainability is, as noted, an inherently temporal concept, and implicitly references an indefinite future. As long as future benefits are discounted at a significant rate—or, really, at any discount rate at all—economics as it is now practiced will be an impediment to a sustainable economy, not an analytic aid toward achieving it.

As these reflections indicate the topics of future inquiry by sustainability theorists are myriad. I will limit my remaining reflections on topics of further inquiry for sustainability theorists in the discipline of philosophy.

The first topic of future philosophical inquiry that comes to mind is the temporal scale of sustainability—as I have characterized it in terms of atuning the human economy to the economy of nature. The spatial scale, as noted, must be global because of the indistinct ontology of the Earth's many ecosystems in comparison with the robust ontology of the Earth itself considered as a systemic unit. So what is a fitting and correlative *temporal* scale for sustainability? It lies somewhere between a scale calibrated in nanoseconds and one calibrated in millions of years, but narrowing that vast range to the proper temporal scale is indeed a topic of future philosophical inquiry.

The second topic of future philosophical inquiry that comes to mind concerns moral ontology. To whom—or perhaps better, to what—is the present generation obligated by way of our widely recognized obligation to "future generations"? Resolving this problem depends not on a definitive resolution to the problem of determining a fitting temporal scale for sustainability, but that topic of future inquiry puts this one in temporal perspective. Whatever the fitting temporal scale of sustainability—a human economy atuned to the economy of nature—we can be sure it extends beyond what we might think of as the personal future.

By "personal future," I mean that, for members of my generation, future generations already exist and my age mates and I are personally acquainted with some members of those future generations. At nearly seventy years of age, my own future is very limited. My son is forty years of age and may well live another forty or fifty years. That forty or fifty year temporal horizon is thus part of my personal future—because I am personally very concerned about the world my son will live in as he approaches my present age. His son, my grandson, is ten years old. He may well live another seventy or eighty years, extending my personal future—members of the future generations with whom I am personally acquainted and for whose welfare I am personally concerned—almost throughout the whole twenty-first century.

I want to help insure that the world that my son and grandson will live in remains a habitable and pleasant world to live in. For that to happen, radical changes in the human economy will be necessary to mitigate and to adapt to global climate change. But radical changes in the human economy will entrain radical changes in human lifestyles, including the reproductive chances and choices of the members of my grandson's generation. That leads to the Parfit paradox. If business goes on as usual, my grandson and other members of his generation will meet, mate, and have one set of children. But if radical changes in the human economy are effected, with attendant lifestyle changes, my grandson and other members of his generation will meet, mate, and have a set of children different from the set of children they would have had had business gone on as usual. So it is impossible for me to be concerned about the individual welfare of my unborn great grandson or -daughter, individually, and that of his or her unborn cohort. For if I succeed in helping make the world of the twenty-second century a habitable and pleasant world in which to live, those individuals who would have existed had business gone on as usual would not exist at all; rather, different individuals would exist in their stead.

Assuming that the temporal scale of sustainability, as I understand it, extends beyond one century into the future, the ontology of future-generations ethics must also be scaled up from one of individual persons to something more proportional to the temporal scale of sustainability, whatever that turns out to be. As noted, the appropriate *objects* of moral considerability for future-generations ethics cannot be yet unborn future *individual* human persons—because their very existence or nonexistence, as individual persons, will depend on what we presently do or leave undone. Then is the appropriate object of moral considerability for future-generations ethics *Homo sapiens*—the human species? I think not. The natural lifespan of a large mammalian species is about a million years—which, I would think exceeds the temporal scale of sustainability. Further, our species may survive an environmental holocaust and hang on with a much-reduced population living in a scarcely habitable and very unpleasant world—if we can trust the vision of post-apocalypse fiction writers and film makers.

No, I think that the appropriate object(s) of moral considerability for future-generations ethics is (are) neither individual human persons nor the human species, but global human civilization. In addition to environmental quality, what I most care about projecting into the future are things like the visual arts, music, poetry, literature, science, and philosophy. Not just the cultural achievements of the past, the *Republic*, the *Bhagavad Gita*, *Hamlet*, the Mona Lisa, but those of the future. But aren't such future achievements of global human civilization equally subject to the Parfit paradox? Yes, they are—if we think of the future achievements of global human civilization in terms of particular artifacts. But if instead we think in terms of the human *activities* themselves—the visual arts, music, poetry, literature, science, and philosophy—then No, moral concern for the future achievements of human civilization is not vitiated by the Parfit paradox.

That's just my opening gambit concerning the moral ontology of future-generations ethics. The question has scarcely been asked, and so a definitive answer is impossible to even broach. This along with many others is an important topic of future inquiry in the field of sustainability philosophy.

CHAPTER THREE

Intrinsic Value and Respect for the Natural Environment

Jack Lee

Apparently humans are either helped or hurt by the conditions of their environment. There ought therefore, to be some form of environmental ethic. Holmes Rolston III claims that an environmental ethic must, illuminate, account for, or ground, appropriate respect for and duty towards the natural environment. Furthermore, it must do this without placing the primary importance on human interests.[1] Given these parameters for formulating such an ethic, simply applying human ethics to environmental affairs would not suffice. How then might we construct such an ethic—an ethic of environment?

Some philosophers (e.g., J. Baird Callicott) believe that what is needed for responding to this challenge is an account of the "intrinsic value"[2] of non-human entities and of nature as a whole. Put more simply, they believe that the natural environment has intrinsic value. Consequently, we should have an appropriate respect for and duty towards the natural environment. This approach however, is rejected by Regan. Although he accepts that an environmental ethic is appropriate, Regan believes that such an ethic *cannot* reasonably be constructed by appealing to "intrinsic value." To explain his demurral, Regan examines four main kinds of theories of intrinsic value. They are: (1) mental-state theories of intrinsic value, (2) states-of-affairs theories of intrinsic value, (3) end-in-itself theories of intrinsic value, and (4) hierarchical ends-in-themselves theories of intrinsic value. After considering each of these, Regan concludes that none can be appealed to in order to construct an environmental ethic. Let us examine each type of theory in more detail.

I.

A classic example of the mental-state theory is hedonism. According to hedonism, pleasure and pleasure *alone* is good in itself. In this case, a mental state of a particular description, and *only* such a mental state, is intrinsically valuable. This hedonist view is seriously challenged by Regan. He writes: "The error of hedonism...lies...in supposing that *only* pleasant experiences...have value of this kind."[3] Indeed, it is reasonable to suppose that if one accepts that an experience of pleasure can be good in itself, then experiences of awe and mystery can likewise be good in themselves. It is false therefore to claim that the only thing of value in valued experiences is the state of pleasure alone.

Even if we adopt a particular mental-state theory of intrinsic value that can overcome the problem of hedonism as outlined above, we will still be faced with yet another difficulty. Mental-state theories of intrinsic value in general *cannot* consistently recognize the intrinsic value of natural (or superorganismic) entities. Regan explains:

> Natural [or superorganismic] entities as such, assuming they lack the requisite psychological capacities, can have no intrinsic value given a mental-state theory of intrinsic value. Lacking a mind, species, populations and ecosystems lack the capacity to have mental states. Thus, [they] lack intrinsic value...[4]

A mental-state theory of intrinsic value therefore, *cannot* even recognize the intrinsic value of some individual organisms, such as gum trees. Since Callicott argues that an environmental ethic must provide for the intrinsic value of both individual organisms and a hierarchy of superorganismic entities (e.g., populations, species, biocoenoses, biomes, and the biosphere),[5] a mental-state theory of intrinsic value does not hold much promise as a tool for fashioning an environmental ethic. Let us therefore move on to state-of-affairs theories of intrinsic value.

G. E. Moore's theory is typical of state-of-affairs theory of intrinsic value. In Moore's view, what is intrinsically valuable can exist independent of a person's mental state. Moore believes that certain states-of-affairs—for example, the beauty of a sunset at a given point in time—have intrinsic value regardless of whether a "mind" is present or not.[6] Unfortunately, neither Moore's view, nor state-of-affairs theories of intrinsic value in general, are well-suited to the development of an environmental ethic either.

Firstly, state-of-affairs theories of intrinsic value cannot illuminate, account for or ground *respect* for the natural environment. Let us assume, like Moore, that beauty is intrinsically valuable. Even if we make this controversial assumption, it will not avail us. Beauty is not in itself an appropriate object of *respect*. With regard to this view, Regan explains: "Granted, one can admire what is beautiful, one can stand in awe of it, one can enjoy or savor or appreciate it, but the idea that one should *respect the beauty in an object* strains our powers of comprehension."[7] It does not, therefore, make any sense to claim that I might, or should, "respect" a beautiful sunset. By the same token, we may conclude: It does not make any sense to claim that I might, or should, "respect" a pleasant experience.

Secondly, state-of-affairs theories of intrinsic value cannot illuminate, account for or ground *duties* to the natural environment. According to state-of affairs theories, there are different kinds of intrinsic values, such as beauty and pleasure.[8] This raises the problem of commensurability between various intrinsic values. It perhaps makes sense to say that pleasure is good in itself. It also perhaps makes sense to say that beauty is good in itself. But, to be specific, it makes no sense to say that the pleasure of drinking a cold Coke on a hot summer day is equal to the beauty of a sunset on the beach at the Gold Coast. These two different kinds of intrinsic values cannot be reasonably treated as if they were commensurate with one another. Given this quandary, state-of-affairs theories of intrinsic value do not afford us any theoretical account of our *duties* to the natural environment.[9]

As both Rolston and Regan assert, an environmental ethic must illuminate, account for, or ground appropriate *respect* for and *duty* towards the natural environment.[10] State-of-affairs theories of intrinsic value cannot however illuminate, account for, or ground appropriate *respect* for and *duty* towards the natural environment. Therefore, state-of-affairs theories of intrinsic value are ill-suited to forming the basis of an environmental ethic.

Let us now turn to end-in-itself theories of intrinsic value. Kant provides an excellent example of these theories. According to Kant, rational autonomous individuals exist as ends-in-themselves. Put another way, these individuals have intrinsic value. Besides, on Kant's view, the intrinsic value, applied to individuals as ends-in-themselves, is a *categorical* concept. That is to say, either individuals exist as ends-in-themselves or they do not. In any case, for Kant, the value is not a matter of degree. Conversely, the theories examined above of mental-state and states-of-affairs suggest that different intrinsic goods can differ in the degree to which they are good in themselves. For example, the hedonist may deny that all pleasures are equally good. Indeed, the hedonist might go on to claim that some pleasures last longer, while others are purer or still, more certain.[11]

Regan suggests a different version of an end-in-itself theory of intrinsic value. Like Kant, he argues that certain individuals have intrinsic value. And he understands the concept of intrinsic value categorically. But whereas Kant limits possession of this value to rational autonomous agents, Regan attributes it to individuals who have a psychological identity over time and thus have an experiential welfare. In short therefore, unlike Kant's theory, Regan's theory acknowledges the intrinsic value of many non-human animals.[12] Paul W. Taylor attempts to take this recognition even further. Taylor proposes that each individual living being—from the simplest unicellular life-forms to the most complex of organisms—has intrinsic value. Moreover, like Kant, he too interprets intrinsic value as categorical concept.[13]

Despite the distinguishing differences between Kant's theory and those of Regan and Taylor, each shares this fundamental similarity: Each theory attributes intrinsic value only to *individuals*—individual rational autonomous agents in Kant's theory, individuals that are the subjects-of-a-life in Regan's theory, and individuals that are teleological centers of life in Taylor's theory.[14] Regardless of the distinctions, the various end-in-itself theories of intrinsic value limits intrinsic value to, and applies only to, *individuals*. And therein lies the concern. To begin with, if an end-in-itself theory of intrinsic value applies and limits intrinsic value only to individuals, then it cannot attribute intrinsic value to natural (or superorganismic) entities, such as populations, species, biocoenoses, biomes, or even the biosphere itself. As outlined above however, an environmental ethic must provide for, or include, the intrinsic value of natural (or superorganismic) entities. Thus, an end-in-itself theory of intrinsic value cannot be employed for constructing an (appropriate) environmental ethic.

Even if an end-in-itself theory of intrinsic value can consistently recognize the intrinsic value of natural (or superorganismic) entities, it fails to ground the differential intrinsic value of what is wild and what is domesticated. It is this consequence that suggests that the theory is inadequate. This point can be expanded on as follows:

> (1) According to an end-in-itself theory of intrinsic value, "intrinsic value" is a *categorical* concept.
> (2) If "intrinsic value" is a *categorical* concept, it must be false to claim that wild and domestic species differ in their intrinsic value.
> (3) However, as Regan (or even Callicott, perhaps) correctly maintains, an environmental ethic must provide for an account of intrinsic value that attributes greater intrinsic value to wild in comparison with domestic species.
> (4) Therefore, an end-in-itself theory of intrinsic value cannot be well-suited to an environmental ethic.[15]

A similar problem arises when we consider the case of endangered species. Rarity appears to confer a special value to a species. Indeed, some environmental philosophers suggest that we should attribute greater intrinsic value to endangered species than we do to those species that are more plentiful. Moreover, those philosophers are prepared to sacrifice, for example, large numbers of plentiful species in order to save endangered forms of life. Their view seems reasonable in terms of an "*environmental* ethic." If however, "intrinsic value" means "end-in-itself," their view must be mistaken.

In the face of the above objections, it seems that we should adopt a hierarchical ends-in-themselves theory of intrinsic value. Regan explains this hierarchical theory as follows: The theory must rank intrinsically valuable entities in some hierarchy of value, from the lower (individual instances of domesticated forms of life, for example, such as this pineapple and that cat), to the next higher level (individual instances of undomesticated forms of life, let us suppose), then to the next, then the next, the next, and so on. That is to say, on this hierarchical theory of intrinsic value, all members of the hierarchy have some intrinsic value; it just happens that some members have more intrinsic value than others. Given this, it is claimed that: each member in the hierarchy exists as end-in-itself and thus normally is not to be treated as a mere means in order to bring about some desirable outcome. This status will change when and if it is necessary to treat them merely as a means to achieving some other higher end-in-itself.

By way of illustration: If the intrinsic value of a balanced, diversified ecosystem is imperiled by an overpopulation of deer, then we will be permitted to use various means, including lethal ones, to control or regulate the population of deer. If however, there were no conflict of intrinsic values between the two parties, then we would not be permitted to cull the population of deer.[16]

Hierarchical ends-in-themselves theories of intrinsic value are superfluous however, because any duty we have with respect to the natural environment that might be illuminated, accounted for, or grounded, in a hierarchy of intrinsic values can be more parsimoniously illuminated, accounted for, or grounded, in a hierarchy of instrumental values. Regan explains:

It is only necessary to say that "lower" forms of life are not "means" to be used unthinkingly or carelessly but are, rather, to be treated as "mere means" *only when this is necessary* in order to protect *higher-ranking members*. On this analysis...it will be wrong to kill deer if their presence does not threaten a "higher good" (for example, the diversity and sustainability of a local habit) but not wrong to do so if it does.[17]

In any case, hierarchical ends-in-themselves theories of intrinsic value, when provided as an explanation of our duties, fail the test of parsimony.

According to the analysis above, it is concluded that "intrinsic value" *cannot* be appealed to in order to construct an environmental ethic.[18]

Since an environmental ethic must illuminate, account for or ground appropriate *respect* for and *duty* towards the natural environment, and given that "intrinsic value" *cannot* be appealed to in order to construct such an ethic, we as philosophers are obliged to explore other approach(es). For the rest of this article, I will endeavor to achieve part of this task. Rather than appealing to "intrinsic value," an alternative account will be briefly offered below of why we should have an appropriate *respect* for (i) individual living things, and (ii) the collective biological entities.

II.

Before embarking on this task, it is first important to distinguish between the idea of "moral consideration" and the idea of "moral significance." While "moral consideration" is used as a tool of judgment, "moral significance" is used to compare assessments of moral weight in cases of conflict. It is the difference, for example, between whether a pine tree deserves any moral consideration and the entirely separate question of whether pine trees deserve more or less consideration than a cat, or a human. Indeed, as Kenneth E. Goodpaster suggests, we should *not* expect that the criterion for having "moral standing" at all will be the same as the criterion for adjudicating the competing claims to priority among beings that merit that standing.[19] The priority issues do have to be dealt with for an operational ethical account, but in the interests of clarity they shall be set aside on this occasion.

Here therefore is the pressing question: What makes a being deserve moral consideration? To this question, Goodpaster offers the following answer:

> (1) Beings capable of being beneficiaries deserve moral consideration from all rational moral agents.
> (2) Beings that have (or can have) interests are capable of being beneficiaries.
> (3) Thus, beings that have (or can have) interests deserve moral consideration from all rational moral agents.[20]

According to this argument, "a being's having (or ability to have) interests" is a sufficient condition for the worth of that being to be considered by all rational moral agents. What then is meant by "a being's having (or ability to have) interests"? Simply stated: If a being is one that can meaningfully be said to be "happy or miserable," or be "well or ill," or even "to flourish or to be injured" and so on, then it has interests. In other words, if a being has objective *goods* (or bads) (such as the flourishing of the being), then it has interests. Indeed, it is in a being's interests that it is happy *or* is well *or* that it flourishes. As a result, it is *good* for this being that it is happy *or* is well *or* that it flourishes. Note that "the flourishing of the being," for example, is quite independent of our evaluations or interests and thus should be regarded as an *objective* state of affairs. The "objective goods principle" may therefore be stated as: *If a being has (or can have) objective goods (or bads), then it deserves moral consideration from all rational moral agents.*

On the basis of the "objective goods principle," I will therefore attempt to show that individual living things (i.e., animals and plants) deserve moral consideration from all rational agents.

Animals are capable of suffering. It is *bad* for my cat for example, if a young boy cuts its tail off just for fun. My poor cat must be feeling extremely painful during and perhaps even after the event. Similarly, if chickens (or bison) are forced to live in cramped, unsuitable conditions for the duration of their lives, it would be *bad* for them. They would be permanently uncomfortable. It is apparent therefore, that animals can have objective bads. Surely then, animals can have objective goods as well. For example, mild winters are *good* for black bears. Given that animals can have objective goods and bads, and given the "objective goods principle," I contend that animals deserve moral consideration from all rational moral agents. This argument can be briefly put as follows:

> (1) If a being has (or can have) objective goods (or bads), then it deserves moral consideration from all rational moral agents (objective goods principle).
> (2) Animals are beings that can have objective goods and bads.
> (3) Therefore, animals deserve moral consideration from all rational moral agents.

As for plants (or any other forms of life), it is likely that they deserve moral consideration from us as well. Plants are vital objects with inherited biological propensities determining their natural growth. Indeed, we are wont to say that certain conditions are *good* (or *bad*) for plants, thereby suggesting that plants can "have" a *good* (or a *bad*). For instance, the warm sunshine is good for my orchid.[21] Moreover, a plant can be said to flourish if it develops those characteristics that are normal to the species to which it belongs in the normal conditions for that species. If it fails to realize such characteristics, then it will be described as "defective," "abnormal," "stunted" and such like.[22] In short, plants can have objective goods or bads. With reference to the "objective goods principle," we can therefore conclude that: plants deserve moral consideration from us.

At a bare minimum it would seem, the mere fact that an entity is a living thing and that it has the capacity for objective goods or bads is sufficient for it to receive moral consideration. This being the case, it follows that because animals and plants deserve moral consideration from us, we as rational moral agents should have appropriate *respect* for them.

Discussion of collective biological entities however, requires a greater degree of subtlety. This is because collective biological entities—colonies, ecosystems and so on—are possessed of life in only the metaphorical sense. They lack those properties typical of living things—reproduction, growth, death and the like. However, collective biological entities *do* have goods (and bads). It does make sense to talk about the conditions in which collective biological entities flourish and thus of their goods. For instance, fresh water and fertilized soil might be essential for a certain ecosystem. Correspondingly, we can meaningfully talk about what is damaging to collective biological entities and hence of their bads. For example, the polluted air around the city tends to be bad for the remnants of the original habitat. In short, collective biological entities can be said to flourish (or to be injured).

Note however that the goods of collective biological entities are not reducible to the goods of their members. John O'Neill illuminates this view:

> The realisation of the good of a colony of ants might in certain circumstances involve the death of most its members. It is not a condition for the flourishing of an individual animal that it be eaten: it often is a condition for the flourishng of the ecosystem of which it is a part...Most members of a species die in early life. This is clearly bad for the individuals involved. But it is...essential to the flourishing of the ecosystems of which they are a part.[23]

Collective biological entities—colonies, ecosystems and so on—can have their own objective goods (or bads), although they are not possessed of their own life. Given collective biological entities can have their own objective goods (or bads) therefore, the "objective goods principle" demands that collective biological entities deserve moral consideration from us. Provided collective biological entities deserve moral consideration from us then, we as rational moral agents should have appropriate *respect* for them. It is clear therefore that the scope of our moral respect should further encompass collective biological entities along with humans, animals and plants.

At this point, someone might challenge the "objective goods principle" thusly:

> Why should any being that has (or can have) objective goods (or bads) deserve moral consideration from all rational moral agents? Furthermore, why should we have an appropriate *respect* for beings that have (or can have) objective goods or bads?

To this challenge, we may respond:

We should recognize individual living things and collective biological entities, as ends-in-themselves. We should have an appropriate *respect* for them. Such a respect for the natural environment is constitutive of a flourishing *human* life. Given that we are rational moral agents, were we to lack respect for our natural environment we would be incomplete as we would lack part of what makes for a flourishing *human* existence. The best human life is one that includes an awareness of and respect for environmental entities with objective goods (or bads).[24]

Endnotes

1. See Holmes Rolston, III, *Environmental Ethics: Duties to and Values in the Natural World* (Philadelphia: Temple University Press, 1988), p. 1.

2. John O'Neill argues that the term "intrinsic value" is used in at least three basic different senses:

(1) Intrinsic value$_1$: Intrinsic value is used as a synonym for non-instrumental value.
(2) Intrinsic value$_2$: Intrinsic value is used to refer to the value an object has solely in virtue of its "intrinsic properties."
(3) Intrinsic value$_3$: Intrinsic value is used as a synonym for "objective value" i.e., value that an object possesses independently of the valuations of valuers.

In this article, what is meant by "intrinsic value" is O'Neill's intrinsic value$_1$. O'Neill contends that to hold an environmental ethic is to hold that non-human beings have intrinsic value$_1$: it is to hold that non-human beings are not simply of value as a means to human ends. On the other hand, Paul W. Taylor offers three kinds of intrinsic value as well:

(1) The immediately good: Any experience or activity of a conscious being which it finds to be enjoyable, satisfying, pleasant, or worthwhile in itself.
(2) The intrinsically valued: An entity is intrinsically valued in this sense only in relation to its being valued in a certain way by some human valuer.
(3) Inherent worth: This is the value something has simply in virtue of the fact that it has a good of its own.

See John O'Neill, "The Varieties of Intrinsic Value," *Monist* 75, No. 2 (1992), pp. 119-120; and Paul W. Taylor, "Are Humans Superior to Animals and Plants?", *Environmental Ethics* 6 (1984), pp. 150-151.

3. Tom Regan, "Does Environmental Ethics Rest on a Mistake?" *Monist* 75, No. 2 (1992), p. 165.

4. Ibid., p. 166.

5. J. Baird Callicott, "Non-Anthropocentric Value Theory and Environmental Ethics," *American Philosophical Quarterly* 21 (1984), p. 299.

6. See chap. III and chap. IV, in G. E. Moore, *Principia Ethica* (Cambridge: Cambridge University Press, 1903); and Tom Regan, ed., *The Elements of Ethics* (Philadelphia: Temple University Press, 1992).

7. Tom Regan, "Does Environmental Ethics Rest on a Mistake?" *Monist* 75, No. 2 (1992), p. 169.

8. Note that while mental-state theories and states-of-affairs theories are conceptually distinct, the latter can include the former. For, as Regan suggests, the idea of a state of affairs is elastic enough to include mental states. Because of this, besides beauty, pleasure can also be said to have intrinsic values. See Ibid., pp. 168-170.

9. See Ibid., p. 170.

10. Holmes Rolston, III, *Environmental Ethics: Duties to and Values in the Natural World* (Philadelphia: Temple University Press, 1988), p. 1. See also Ibid., pp. 161-162.

11. Likewise, for Moore, a beautiful sunset of which no one is aware has some intrinsic value, while the complex whole consisting of this same sunset plus someone's admiration and enjoyment has yet more intrinsic value. See Tom Regan, "Does Environmental Ethics Rest on a Mistake?" *Monist* 75, No. 2 (1992), pp. 172-173.

12. See Tom Regan, *The Case for Animal Rights* (Berkeley, CA: University of California Press, 1983); and Tom Regan, "Empty Cages: Animals Rights and Vivisection," in Tony Gilland, ed., *Animal Experimentation: Good or Bad?* (London: Hodder & Stoughton, 2002), pp. 19-36.

13. See Paul W. Taylor, *Respect for Nature: A Theory of Environmental Ethics* (Princeton, N.J.: Princeton University Press, 1986).

14. See Tom Regan, "Does Environmental Ethics Rest on a Mistake?" *Monist* 75, No. 2 (1992), pp. 173-174.

15. See Ibid., pp. 162, 175-176; and J. Baird Callicott, "Non-Anthropocentric Value Theory and Environmental Ethics," *American Philosophical Quarterly* 21 (1984), pp. 299-309. Note that "wild and domestic species differ in their intrinsic value" in (2) can be interpreted *either* individually *or* collectively (or holistically).

16. See Tom Regan, "Does Environmental Ethics Rest on a Mistake?" *Monist* 75, No. 2 (1992), pp. 177-179.

17. Tom Regan, "Does Environmental Ethics Rest on a Mistake?" *Monist* 75, No. 2 (1992), p. 178.

18. See Ibid., pp. 179-180. Regan argues that even if there is a fifth, sixth or some other possible theory of intrinsic value, it will succumb to one or other of objections brought against the four considered above:

    (1) The supposed theory cannot be either a mental-state or state-of-affairs theory for the reasons given in the above.
    (2) Whatever form such a theory might take, it will have to imply *either* that all intrinsically valuable entities are equal *or* that they are not.
    (3) If the former, then such a theory will not be able to account for the difference in intrinsic value that is supposed to hold between what is wild and what is domestic.
    (4) If the latter, then there simply will be no parsimonious reason for supposing that "lower" members of the hierarchy have "some" intrinsic value in the first place.

19. See Kenneth E. Goodpaster, "On Being Morally Considerable," in Michael E. Zimmerman, J. Baird Callicott, George Sessions, Karen J. Warren, and John Clark, eds.,

Environmental Philosophy: From Animal Rights to Radical Ecology, 3rd (New Jersey: Prentice-Hall, 2001), p. 59.

20. Ibid., p. 64.

21. On the other hand, mere things (e.g., a stone) cannot have any goods (or bads). They have no "well-being" to be sought or acknowledged by rational moral agents.

22. See Ibid., p. 65; and John O'Neill, "The Varieties of Intrinsic Value," *Monist* 75, No. 2 (1992), pp. 129-130.

23. John O'Neill, "The Varieties of Intrinsic Value," *Monist* 75, No. 2 (1992), p. 131.

24. See Ibid., pp. 132-133.

CHAPTER FOUR

The Land Ethic and Gleason's Individualistic Concept of Plant Association

Allen Yu

Introduction

The land ethic, primarily functioning as an ethical approach to environmental problems identified in Aldo Leopold's classic book, *A Sand County Almanac*, is one of the foremost pioneering and inspiring attempts to formulate environmental ethics. Leopold's land ethic is to a large extent based on the *community* concept in ecology. For instance, the premise of the land ethic is that the individual is a member of the land-community of interdependent parts. And the maxim of the land ethic also draws attention to the need to preserve the integrity, stability, and beauty of the land community.[1] The community concept[2], representing a holistic paradigm holding that biotic communities are integrated and stable, has become problematic since the late 1940s, when a number of individualistic ecologists, such as H. A. Gleason, started arguing that plant associations do exist, but that these eco-groupings are too loosely related to constitute an integrated and stable community. Against this background, the aim of this paper is threefold: (1) to briefly outline Gleason's individualistic argument and how it would affect the land ethic; (2) to argue that Gleason's individualism raises, at least, a threefold problem for the land ethic, including (i) how to incorporate ever-developing sciences within the land ethic, (ii) how to engender duties from ever-changing communities, and (iii) how to modify the maxim of the land ethic; and (3) to argue that Leopold's adaptive scientific epistemology, the affirmation of eco-relationships, and a new formulation of the maxim can provide replies to the problems raised by Gleason's individualism.

Gleason's Individualistic Concept of the Plant Association

Gleason, an American ecologist, published "The Individualistic Concept of the Plant Association," in 1926 and again in 1939[3]. These essays were largely ignored in the first half of the twentieth century. However, after World War II, Gleason's individualism, in virtue of some potent confirmation from the field studies on prairies, gained growing recognition and became a leading trend in ecology.[4] Some even argue that this is a paradigm shift from holism to individualism in ecology.[5] This background shows that Gleason's individualistic theory played an influential role in inspiring ecologists to question the community theory.[6] A quick survey of Gleason's individualistic concept of plant association can be found in his 1939 paper, in which he concisely presented his individualistic argument with seven theses. They are as follows:

1. The ordinary processes of migration bring the reproductive bodies of a single plant or a species of plant into many places.
2. The ordinary processes of migration bring the reproductive bodies of various plants into the same place.
3. Of the various species which reach one spot of ground, the local environment determines which may live, depending on the individual physiological demands of each species separately.
4. On every spot of ground, the environment varies in time, and consequently the vegetation varies in time.
5. At any given time, the environment varies in space, and consequently vegetation varies in space.
6. A piece of vegetation which maintains a reasonable degree of homogeneity over an appreciable area and a reasonable permanence over a considerable time may be designated as a unit community. Within such an area and during such a period similarity in environmental selection tends toward similarity in vegetation.
7. Since every community varies in structure, and since

no two communities are precisely alike, or have genetic or dynamic connection, a precisely logical classification of communities is not possible.[7]

A summary of this argument is as follows:

(1) The formation of vegetation units is due to the interaction of individual species' migration and the environmental selection. (Theses 1, 2, and 3)
(2) The interaction of individual species' migration and the environmental selection varies in time and space. (Theses 4 and 5)

Therefore, (3) the formation of vegetation units, though exists, varies in time and space. (Theses 6 and 7)

According to this argument, a possible impact of Gleason's individualistic community model on Leopold's ecological account is his reconsideration of community integrity and stability. According to the land ethic, when addressing the community concept, Leopold explicitly stressed five ideas: interdependency (e.g. food and energy interdependency), health (e.g. the health of the land, understood by Leopold as the capacity for self-renewal), integrity, stability, and beauty. Compared with Gleason's individualistic argument, we can discern, firstly, that the ideas of health and beauty are not addressed in Gleason's argument, and, secondly, that the other three ideas are related to Gleason's individualism in different ways. On the one hand, Leopold's stress of interdependency is positively recognized by Gleason's emphasis of the interaction between each species' migration and the environment's selection. On the other hand, Leopold's account of integrity and stability is likely to be, if not rejected, reformulated by Gleason's statement of time and space variation within communities. Thus, a way to show how Gleason's individualism constitutes a counter-example to Leopold's ecological account, is to consider how Gleason's individualism is shaped by his account of integrity and stability.

I first show how Gleason reformulated integrity. Despite the fact that integrity has been variously defined by various ecologists who have different concerns[8], a general definition of community integrity can refer to the maintenance of the community structure as complete or undivided. To argue against the community structure as being complete or undivided, Gleason invited us to think of how component species appear and disappear in a community. Imagining a simplified community originally consisting of species A, B, and C, and later replaced by species X, Y, and Z. Gleason illustrated this transition process as follows:

> ...species X, Y, and Z must have been derived from some neighboring area or areas.... In this neighboring area species X, Y, and Z may or may not have been living together. *In these original stations the environment was variable, their seed production was variable, and the migration of these seeds was controlled largely by those inexplicable factors which we call chance....* The plants of the earlier association, which we may designate A, B, and C, eventually reached the end of their time-period and disappeared. *Their disappearance may or may not have been simultaneous.* The particular environmental changes which facilitated the entrance of X, Y, and Z may not have been the changes which caused the loss of A, B, and C. The former may have entered first, so that at one time the association consisted of all six species, or entry and disappearance may have been more or less serial, so that the association passed through successive stages of A, B, C; X, B, C; X, Y, C; and finally X, Y, Z.[9]

The formation of communities, for Gleason, is dependent upon how the individual physiological demands of each species separately interact with the environment (see Thesis 3). The community structure is merely a contingent combination of variable species' physiological demands and the variable environment. Therefore, each component species, such as A, B, C, X, Y, and Z, is not inseparable as an integrated entity. As the environment and species' physiological nature vary over time and space, the community structure will also change. The integrity of the community structure, as Gleason suggested, is only a result of chance existing to a reasonable extent in time and space.

After showing Gleason's reconsideration of integrity, I now turn to stability. It too has been defined variously.[10] Nevertheless, before the concept of stability became diverse, it was traditionally related to the balance or equilibrium of nature[11], and widely utilized by ecologists to "characterize a stage of a community perpetuating itself barring disturbance."[12] One conspicuous problem with the stability concept, noted by Gleason and many other ecologists, is that stability does not make clear the scale or the duration under consideration. Gleason said:

> If our lives were measured by days instead of years, we can imagine an ecologist saying, "I have seen three generations of Galinsoga on this spot of ground. Evidently we are dealing with a stable environment and Galinsoga will live here forever." He would be wrong. With our knowledge of vegetational conditions actually extending over about three centuries, we now say, "Oaks have occupied this spot of ground for three hundred years. Evidently we have here a stable environment and the oak will live here forever." If our lives were seventy centuries instead of seventy years, would we not see that we were again wrong?[13]

One of the main difficulties of assessing community stability is that we lack methods lasting enough to observe different time-scales of environmental variations. For example, Gleason claimed that there are two classes of environmental variations which are unpredictable and unquantifiable. The first is fluctuation, "illustrated by our irregular alternation of cold and warm, of dry and wet years, of late and early seasons."[14] Because of the irregularity, this kind of environmental variations is unpredictable. The second is "cumulative environmental changes which progress over a period of years or centuries or ages,"[15] such as the silting up of a pond. Because it often progresses slowly over a long period of time, this kind of environmental variations is difficult to measure. Consequently, without proper means of predicting and measuring environmental variations in different time-scales, using the concept of stability to describe communities, as Gleason warned, is at the risk of being wrong.

To sum up, this section exhibits Gleason's individualistic argument and elucidates how the individualistic community concept would affect the concepts of integrity and stability. Based on this, I will go on to show how Gleason's individualism raises a threefold problem for the land ethic.

The Problems for the Land Ethic Raised by Gleason's Individualism

I now turn to reply to some—but by no means all—of the salient problems with the role of ecology in the land ethic. My discussion of the problems raised by Gleason's individualism is threefold: (i) how to incorporate ever-developing sciences within the land ethic, (ii) how to engender duties from ever-changing communities, and (iii) how to modify the maxim of the land ethic. I will further develop this threefold problem in order.

To begin with (i). How to incorporate ever-developing sciences is always a difficult problem and one encountered by many philosophers. As Robert Kirkman points out, "many environmental philosophers relate to the natural sciences as they relate to the philosophical tradition.... Selective science, like selective philosophy, has its dangers."[16] One reason why relying on the sciences is dangerous is because "however useful [the sciences] might be, the tentative truths of the sciences rest within a much larger and deeper uncertainty."[17] Based on this, Philip Cafaro claims that one of Kirkman's meritorious points is that "philosophers who pin their ethics too closely to particular scientific theories are at the mercy of future changes in science."[18] The entanglement of Leopold's land ethic with ecology is a typical example. In light of the land ethic, in order to extend moral consideration toward nature, Leopold utilized ecology as an intellectual subject guiding people to perceive the eco-relationships, such as the membership of the land community of interdependent parts. Based on these eco-relationships, Leopold proposed the premise and maxim of the land ethic, as follows:

1. The individual is a member of the land-community of interdependent parts.

2. A thing is right when it tends to preserve the integrity, stability, and beauty of the biotic community. It is wrong when it tends otherwise.[19]

However, because Leopold pinned his ecological explanation too closely to the holistic paradigm of the community concept, once the paradigm shifted to individualism, Leopold's ecological account, including the eco-relationships and the premise and maxim of the land ethic, is at the mercy of the individualistic turn in ecology. Therefore, a tight entanglement with paradigm-shifting ecology will require the land ethic, if defensible, to have a good strategy to resist the uncertain development of ecology.[20]

To further address the effects of the uncertain impact of ecology on the land ethic I will go to consider (ii) and (iii). In developing the land ethic, Leopold utilized two ecological concepts, the biotic community and the land pyramid, to demonstrate eco-relationships, such as the human membership of the land community. It was on the basis of these eco-relationships that Leopold asserted the premise and maxim of the land ethic. Moreover, because Leopold embraced the holistic model of the community concept, he viewed communities as integrated and stable entities, and thus the eco-relationships among communities were regarded as integrated and stable as well. However, if communities are not integrated and stable, then the eco-relationships among them will not be integrated and stable as well. Consequently, ecological individualism raises a question of whether the premise and maxim of the land ethic are defensible if the eco-relationships are not integrated and stable.

The questionable premise and maxim will thus lead the land ethic to confront the questions (ii) and (iii). Regarding (ii), the community concept in the premise of the land ethic is conceived as integrated and stable. And based on the premise, the duties and obligations established by the land ethic are derived from the integrated and stable land community. However, if the community paradigm is changed to individualism, then the land ethic will encounter a problem of how to generate duties and obligations if the land community is not integrated and stable but ever-changing.

Likewise, regarding (iii), if the land community is not conceived as integrated and stable, then the maxim of the land ethic emphasizing integrity and stability must be, if not abandoned, at least modified. Therefore, the question of how to properly modify the maxim of the land ethic must be seriously considered.

So far we know that Leopold's ecological account, at least, raises a threefold problem. Nevertheless, these three aspects of the problem all converge toward one concern: "from whence can we make the land ethic adaptive to the individualistic development of ecology?" A resolution of this concern with its three problematic aspects, I propose, should include Leopold's adaptive scientific epistemology, the affirmation of the eco-relationships, and a new formulation of the maxim of the land ethic.

A Reply to the Problems Raised by Gleason's Individualism

I begin with the adaptive scientific epistemology, which is proposed to resolve the question of how to incorporate ever-developing sciences in the land ethic. When Callicott defends the diversity-stability hypothesis in the land ethic, he says:

> The science that informed Leopold's land ethic may be out of date, but Leopold's scientific epistemology seems to be much more advanced than that of some of the contemporary deconstructive ecologists.[21]

The main reason why Callicott exalts Leopold's scientific epistemology is because he notes Leopold's caution in asserting the diversity-stability hypothesis. For example, Leopold averred that "these creatures are members of the biotic community, and if (as I believe) its stability depends on its integrity, they are entitled to continuance."[22] And according to this statement, Callicott observes:

> Leopold's scientific scruples are evident in the parenthetical phrase "as I believe." He does not, that is, state the dependence of stability on diversity (or integrity, that is the presence of the characteristic species of a biotic community

in their characteristic numbers) as a well-established fact, but as his own well-educated opinion.[23]

I share a basic agreement with Callicott's point, but I will go on to extend this point since there is a clue suggesting that Leopold's caution applies not only to the diversity-stability hypothesis, but also to ecological science generally.

This clue is Leopold's distinction between ecology and ecological situations. As Leopold said, "an ethic may be regarded as a mode of guidance for meeting *ecological situations*."[24] Ecology is a branch of science which studies ecological facts, but the complexity of ecological situations has not been entirely disclosed by ecology. Ecological situations are, as Leopold said, "so new or intricate, or involving such deferred reactions, that the path of social expediency is not discernible to the average individual."[25] This points out that the land ethic aims at disclosing intricate ecological situations to people, and ecology is merely a means to proffer information for understanding these situations. However, Leopold was aware that this tool cannot guarantee the achievement of the aim because ecological situations are too complex to be understood by ecological scientists. For instance, when speaking of the biotic mechanism of communities, he said:

> The ordinary citizen today assumes that science knows what makes the community clock tick; the scientist is equally sure that he does not. He knows that the biotic mechanism is so complex that its workings may never be fully understood.[26]

To point out the inadequacy of science to deal with this complexity is the first step. To finalize an adaptive scientific epistemology needs the second step, which is to reduce the impact caused by the inadequacy of science. To do so Leopold needed two things: a distinction and an alternative. The distinction needed is between the land ethic and ecology. As mentioned above, the aim of the land ethic is to provide guidance for understanding ecological situations, and ecology provides the means to achieve this aim. Hence, the land ethic and ecology are distinct and the linkage between them lies in ecological situations. However, although the land ethic and ecology are connected by ecological situations, they utilize ecological situations with different purposes. The land ethic is an ethical theory in pursuit of norms, and ecology is a scientific subject in pursuit of the truth. Thus, on the surface, the land ethic aims at guiding people to meet ecological situations, but in fact its ultimate goal is not to disclose truth, but to lead people to act appropriately. How can ecological situations generate normative force? The answer to this, according to the previous story, is because of the eco-relationships which exist within ecological situations. It is in virtue of eco-relationships that people can extend their moral sentiments toward nature. Finally, through this distinction, I can clarify two important points. Firstly, ecology is merely a means, and this suggests that there can be other means. Once ecology fails to provide guidance, we can provide other means. Secondly, to assess whether ecology fails or not requires a precise understanding of the aim of the land ethic. According to the distinction, I suggest that the ultimate goal of the land ethic is to provide guidance to meet eco-relationships, on the basis of which we can accurately determine whether ecology, after the individualistic turn, can achieve this goal or not.

I now turn to the alternatives. One may ask whether Leopold provided alternatives, and the answer to this, I suggest, is "yes." There are two alternatives. One is in "The Land Ethic," and there is another mainly in "Conservation Esthetic." I first deal with the former, which is "animal instincts." Said Leopold:

> An ethic may be regarded as a mode of guidance for meeting ecological situations so new or intricate, or involving such deferred reactions, that the path of social expediency is not discernible to the average individual. Animal instincts are modes of guidance for the individual in meeting such situations. Ethics are possibly a kind of community instinct in-the-making.[27]

As mentioned above, Leopold's intention, in the land ethic, is to establish normative principles, so when he referred to ecological situations and animal instincts, these two things should be ethically understood as the eco-relationships and community instincts. Hence, immediately after this passage, Leopold continued to explain how community instincts can guide eco-relationships. He said:

> All ethics so far evolved rest upon a single premise: that the individual is a member of a community of interdependent parts. His instincts prompt him to compete for his place in that community, but his ethics prompt him also to co-operate (perhaps in order that there may be a place to compete for).[28]

In short, community instincts guide people to find their places in communities via competition and cooperation. This echoes Leopold's account of the eco-relationships. For the place of people in communities, in other words, is humans' community membership; competition and cooperation, constitute interdependency. For instance, in food-chains and food-webs, species competition and cooperation form the food-interdependency. Therefore, besides ecology, Leopold's community instincts provide the first alternative way of guiding people to appreciate the eco-relationships among ecological situations.

The second alternative is "perception." By investigating Leopold's earlier works, such as "Some Fundamentals of Conservation in the Southwest," Norton finds that perception plays a significant role in Leopold's thought. For example, in the final section, "Conservation as a Moral Issue," Leopold said: "Possibly, in our intuitive perceptions, which may be truer than our science and less impeded by our words than our philosophies, we realize the indivisibility of the earth."[29] Here Leopold explicitly expressed his inclination to intuitive perceptions since they are truer than science and less impeded than philosophy. Although some scholars may suggest that significant ideas here were later abandoned by Leopold[30], his faith in the veracity of perception, at least, is an exception. It is preserved in "Conservation Esthetic." When speaking of various components of outdoor recreation, Leopold said:

> The perception of the natural processes by which the land and the living things upon it have achieved their characteristic forms (evolution) and by which they maintain their existence (ecology). That thing called 'natural study.'[31]

Here Leopold affirmed that perception can lead people to naturally study evolutionary and ecological facts. Moreover, as he originally thought that perception is truer than science, Leopold seemed to suggest that perception is more fundamental than ecological science. He said:

> Ecological science has wrought a change in the mental eye. It has disclosed origins and functions for what to Boone were only facts…. The incredible intricacies of the plant and animal community…were as invisible and incomprehensible to Daniel Boone as they are today to Mr. Babbitt…. [But] Let no man jump to the conclusion that Babbitt must take his Ph.D. in ecology before he can 'see' his country. On the contrary, the Ph.D. may become as callous as an undertaker to the mysteries at which he officiates. Like all real treasures of the mind, perception can be split into infinitely small fractions without losing its quality…. Perception, in short, cannot be purchased with either learned degree or dollars; it grows at home as well as abroad, and he who has a little may use it to as good advantage as he who has much.[32]

Perception is surely a vital supplement to ecological science since it provides guidance within ecological situations. However, the merit of perception is more than its being an alternative to ecology. Perception should be regarded as a more fundamental guide in ecological situations than ecology. The land ethic aims to provide guidance for people to appreciate the eco-relationships within nature, and thus generate "love, respect, and admiration for land."[33] To engage with nature, such as through the perception of natural processes, is more straightforward and more basic than participating in academic discussions. As Leopold suggested, the learning process of ecological knowledge should not only take place in lectures about ecology, but also can and should occur in a wide range of circumstances, including our direct engagement with nature.[34]

To sum up, the awareness of the inadequacy of ecological science and the two alternatives constitute an understanding of Leopold's adaptive scientific epistemology. Therefore, whatever the development of ecology turns out to be, community instincts and perception are vital components in people's natural talents to appreciate nature.

After replying to (i), I now turn to (ii), which is the question of how to engender duties from ever-changing communities. In his reply to this question, Callicott makes two points. Firstly, he asserts the existence of interdependency by quoting Donald Worster's "principle of interdependency." Worster says: "No organism or species of organism has any chance of surviving without the aid of others."[35] Secondly, Callicott utilizes the multiple and dynamic composition of human communities, such as Stevens Points, to provide an analogy with ever-changing biotic communities, and concludes:

> ...if paradigmatic human communities are sufficiently robust to engender civic duties and obligations both to fellow members and to such communities per se, then biotic communities, which are not less robust than paradigmatic human communities, are, by parity of reasoning, also sufficiently robust to engender analogous environmental duties and obligations.[36]

I share a basic agreement with Callicott's points, especially his (and Worster's) affirmation of interdependency, and I will further develop them according to my own interpretation of the land ethic. Callicott is insightful in analogizing loosely grouping human communities to poorly bounded biotic communities, but this analogy is too general to fully explain why human and biotic communities are robust enough to generate analogous duties. The decisive reason why human and biotic communities can generate analogous duties, I suggest, is because they both possess analogous interdependency. One of Leopold's crucial strategies for establishing the land ethic is to explain an ethic as a process of evolution. In "Outlook," he said: "I have purposely presented the land ethic as a product of social evolution because nothing so important as an ethic is ever 'written.'"[37] The purpose of Leopold's strategy is to provide a way of extending ethics to include nature. The extension of ethics can be compared to the extension of railway, in which humanity is like a train, which needs two things to reach a new destination: new tracks and fuel. Track represents the relationship responsible for guiding people across the old boundary towards a new destination. Fuel stands for motivation which is responsible for generating energy to enable the train (humans) to reach the new destination. Therefore, I suggest that two fundamentally important elements for the extension of ethics are relationship (moral track) and motivation (moral fuel). In the case of the land ethic, these two elements are the eco-relationships and sentiments (e.g. love). This framework illuminates one very important point. The duties of the land ethic, as an extended ethic, are generated from the reciprocity of relationships and sentiments. Hence, the merit of Worster's principle of interdependency lies in its affirmation enabling the eco-relationships to be ethically adaptive to ecological individualism. Although communities may not be integrated and stable, once the interdependent relationships are affirmed, as Worster and Callicott claim, the duties of the land ethic are basically secured since in virtue of the eco-relationships people can extend their moral sentiments to nature.

Notwithstanding, that to say that the duties of the land ethic are basically secured does not imply that duties from individualistic communities are the same as duties from holistic communities. Different relationships generate different duties. For example, duties derived from parental-filial relationships must differ from duties derived from friendship. Thus, although the duties from ever-changing communities can be substantiated, the land ethic still needs to revise or even give up some duties which are too closely tied to the holistic community model, such as the duty of preserving the integrity and stability of the land community. This will lead us to (iii), which is the question of how to modify the maxim of the land ethic.

Before addressing (iii), it is important to note one thing. The reason why the maxim the land ethic requires modification is not because integrity and stability are entirely abandoned, but because of the vagueness of these two concepts. Ecological individualism, though powerful, does not completely obliterate the concepts of integrity and stability. For example, McIntosh observes that "the term integrity was absent from most scientific ecological literature until the 1960s, but has increased since then."[38] Also, Shrader-Frechette and McCoy's review of Lewontin, Holling, and Orians suggests that "in the late 1960s and early 1970s a wide range of concepts, with a variety of applications, could be found under the umbrella of 'ecological stability'."[39] Their observations demonstrate that integrity and stability do not disappear from the discourse of ecology after individualism became popular in the 1950s. However, after the challenge of individualism, ecologists who seek a precise definition indeed find variously defined integrity and stability are vague. Therefore, Leopold's loose usage of these vague concepts will make his maxim too equivocal to be a reliable guide.

One way to avoid the vagueness is to abandon the original maxim and reformulate a new one. As Callicott observes, "the individualistic-dynamic paradigm in deconstructive community ecology seems to undercut two [integrity and stability] out of three of the land ethic's cardinal values."[40] As a result, he invokes "scale" as a new normative key, and reformulates the maxim as such: "a thing is right when it tends to disturb the biotic community only at normal spatial and temporal scales. It is wrong when it tends otherwise."[41] Callicott's attempted reformulation is admirable, but his new maxim can be more ethically straightforward if his new normative key—scale—is substituted for the capacity of the land for self-renewal.

The reason why the capacity of the land for self-renewal is more ethically straightforward can be illustrated by questioning Callicott's supporting claims. The first question is that the supporting claims to which Callicott appeals do not explain why being "naturally abnormal" is equal to being "land-ethically wrong." For example, in the case of mass extinction, Callicott says that "normally, speciation out-paces extinction," so "the current *rate* of extinction is wildly abnormal."[42] Here he clearly explains that the reason why the current rate of anthropogenic extinction is abnormal is because extinction outpaces speciation. By the same token, in the case of global warming, he also claims that "we may be causing a big increase of temperature at an unprecedented rate,"[43] but without any additional argument, he further avers that "that's what's land-ethically wrong with anthropogenic global warming."[44] With these two cases, Callicott, at best, demonstrates that anthropogenic perturbations are naturally abnormal or unprecedented, but the reason why being naturally abnormal is equivalent to being ethically wrong is obscure.

In reply to the first question, Callicott may appeal to other environmental philosophers, such as Leopold, Stewart Pickett, and Richard Ostfeld. However, this appeal remains questionable because scale, in light to those philosophers' accounts, is not ethically straightforward. For example, in "The Land Pyramid," Leopold wrote that "evolutionary changes…are usually slow and local. Man's invention of tools has enabled him to make changes of unprecedented violence, rapidity and scope."[45] Leopold's statement here, as the above examples, only exhibits that the measure of rate and scope can "ecologically" judge whether human-made changes are unprecedented. To know Leopold's ethical measure needs to review Leopold's complete argument. He said:

> When a change occurs in one part of the circuit, many other parts must adjust themselves to it. *Change does not necessarily obstruct or divert the flow of energy*; evolution is a long series of self-induced changes, the new result of which has been to elaborate the flow mechanism and to lengthen the circuit. Evolutionary changes, however, are usually slow and local. Man's invention of tools has enabled him to make changes of unprecedented violence, rapidity, and scope.[46]

In light of this statement, Leopold seemed to argue that rapid and widespread human-caused changes can be land-ethically wrong if they obstruct or divert the self-adjustable circuit of the land, such as energy flows and food chains. The land community does not merely stand for a place all natural entities reside in, but also represents a series of mechanisms (e.g. energy flows and food chains) which all entities rely on. To damage the self-renewable capacity of the land is land-ethically unacceptable since it will fatally threaten all of natural entities and the land community to which we belong. Similarly, Pickett and Ostfeld note that:

> ...an inference [any human-caused flux is justifiable] is wrong because *the flux in the natural world has severe limits*... Two characteristics of human-induced flux would suggest that it would be excessive: fast rate and large extent.[47]

Their note implies that fast and far-reaching human-induced flux is naturally excessive, but the reason why excessive human-generated flux is ethically wrong is because it goes beyond the limit of natural flux. That is, in Leopold's words, to obstruct or divert the flow of energy. Therefore, according to Callicott's supporting examples and quotes, the self-renewal of the land community seems to be more ethically straightforward than scale, especially in the case of the land ethic. Based on this, a new reformulation of the maxim of the land ethic, I suggest, can be: *a thing is right when it tends to preserve the capacity for self-renewal of the land community. It is wrong when it tends otherwise.*

This new summary moral maxim is advantageous because it echoes other significant concepts in the land ethic, such as sustainability and land health. For example, in addition to describing the land as self-adjustable, Leopold also mentioned that the circuit of the land is "sustained."[48] Also, in the beginning of "Land Health and the A-B Cleavage," Leopold said:

A land ethic, then, reflects the existence of an ecological conscience, and this in turn reflects a conviction of individual responsibility for the health of the land. Health is the capacity of the land for self-renewal. Conservation is our effort to understand and preserve this capacity.[49]

Because of these echoes, the new maxim is well-matched in the land ethic. However, the new maxim can nevertheless be criticized since it also suffers from vagueness. One apparent problem of the new maxim is how to define the self-renewal of the land. In the face of the imperfect understanding of nature, human beings, at this point, indeed are unable to define the self-renewal of the land without some vagueness. However, unlike the vagueness of integrity and stability, the ambiguity of the self-renewal of the land is identified by Leopold. He said: "biotas seem to differ in their capacity to sustain violent conversion."[50] Due to this observation, he suggested that:

> Whatever may be the equation for men and land, it is improbable that we as yet know all its terms. Recent discoveries in mineral and vitamin nutrition reveal unsuspected dependencies in the up-circuit: incredibly minute quantities of certain substances determine the value of soils to plants, of plants to animals. What of the down-circuit? What of the vanishing species, the preservation of which we now regard as an esthetic luxury? They helped build the soil; in which unsuspected ways may they be essential to its maintenance? Professor Weaver proposes that we use prairie flowers to re-flocculate the wasting soils of the dust bowl; who knows what purpose cranes and condors, otters and grizzlies may some day be used?[51]

In the face of the imperfect understanding of the self-adjustment of the land, not only can we make effort to understand this capacity, we also should be cautious since the effect of our behaviours on nature could be unintended harm. Therefore, although the new maxim also suffers from the vagueness, it is still defensible because Leopold already provided a resolution of the vagueness.

Conclusion

In conclusion, I have briefly showed Gleason's individualistic concept of plant association, and argued how his individualism affect Leopold's land ethic. Based on this, we show that although ecology is an important aid, its uncertain development, such as the individualist turn, poses a threat to the land ethic. In reply to the three problematic aspects of Leopold's ecological account, I suggest that Leopold's adaptive scientific epistemology, the affirmation of eco-interdependency, and a new formulation of the maxim can adapt the land ethic to the challenge of ecological individualism.

Bibliography

Cafaro, P. 2004. "Review of *Skeptical Environmentalism: The Limits of Philosophy and Science.*" *Environmental Ethics* 26: 101-04.

Cairns, J., Jr. 1975. "Quantification of Biological Integrity." in *The Integrity of Water: Proceedings of a Symposium, March 10-12, 1975, Washington, D.C.* Washington, D.C.: Office of Water and Hazardous Materials and U.S. Environmental Protection Agency.

Callicott, J. B. 1989. *In Defense of the Land Ethic: Essays in Environmental Philosophy.* Albany, N.Y.: State University of New York Press.

Callicott, J. B. 1999. *Beyond the Land Ethic: More Essays in Environmental Philosophy.* Albany, N.Y.: State University of New York Press.

Fiedler, P. L., P. S. White, and R. A. Liedy. 1997. "The Paradigm Shift in Ecology and Its Implications for Conservation." in S. T. A. Pickett, R. S. Ostfeld, M. Shachak and G. E. Likens (ed.), *The Ecological Basis of Conservation: Heterogeneity, Ecosystems and Biodiversity.* New York: Chapman and Hall, pp. 83-92.

Flader, S. L. 1974. *Thinking Like a Mountain.* Lincoln: University of Nebraska Press.

Gleason, H. A. 1926. "The Individualistic Concept of the Plant Association." *Bulletin of the Torrey Botanical Club* 53: 7-26.

Gleason, H. A. 1939. "The Individualistic Concept of the Plant Association." *American Midland Naturalist* 21: 92-110.

Grimm, V., and C. Wissel. 1997. "Babel, or the Ecological Stability Discussion: An Inventory and Analysis of Terminology and a Guide for Avoiding Confusion." *Oecologia* 109: 323-34.

Karr, J. R. 1997. "Measuring Biological Condition, Protecting Biological Integrity." in G.K. Meffe and C.R. Carroll (ed.), *Principles of Conservation Biology*. Sunderland, MA: Sinauer, pp. 283-85.

Kirkman, R. 2002. *Skeptical Environmentalism: The Limits of Philosophy and Science*. Bloomington: Indiana University Press.

Leopold, A. c1949. *A Sand County Almanac, and Sketches Here and There*. New York: Oxford University Press.

Leopold, A. 1979. "Some Fundamentals of Conservation in the Southwest." *Environmental Ethics* 1: 131-41

McIntosh, R. P. 1958. "Plant Communities." *Science* 231: 115-20.

McIntosh, R. P. 1967. "The Continuum Concept of Vegetation." *Botanical Review* 33: 130-87.

McIntosh, R. P. 1975. "H. A. Gleason-"Individualistic Ecologist" 1882-1975: His Contributions to Ecological Theory." *Bulletin of the Torrey Botanical Club* 21: 253-73.

McIntosh, R. P. 1980. "The Background and Some Current Problems of Theoretical Ecology." *Synthese* 43: 195-225.

McIntosh, R. P. 1985 *The Background of Ecology: Concept and Theory*. Cambridge: Cambridge University Press.

McIntosh, R. P. 2002. "Ecological Science, Philosophy, and Ecological Ethics." in Jim Hill and Wayne Ouderkirk (ed.), *Land, Value, Community: Callicott and Environmental Philosophy*. Albany: State University of New York, pp.59-84.

Moravec, J. 1989. "Influences of the Individualistic Concept of Vegetation on Syntaxonomy." *Vegetatio* 81, 29-39.

Nicolson, M. 1990. "Henry Allan Gleason and the Individualistic Hypothesis: The Structure of a Botanist's Career." *Botanical Review* 56: 91-161.

Norton, B. G. 2005. *Sustainability: A Philosophy of Adaptive Ecosystem Management*. Chicago: The University of Chicago Press.

Pickett, S. T. A. and R. S. Ostfeld. 1995. "The Shifting Paradigm in Ecology." in R. L. Knight and S. F. Bates (ed.), *A New Century for Natural Resources Management*. Washington, D.C.: Island Press, pp. 261-79.

Shrader-Frechette, K.S., and E.D. McCoy. 1993. *Method in Ecology: Strategies for Conservation*. Cambridge: Cambridge University Press.

Whittaker, R. H. 1962. "Classification of Natural Communities." *Botanical Review* 28: 1-239.

Worster, D. 1994. *Nature's Economy: The Roots of Ecology*. Garden City, New York: Anchor Books.

Notes

1 It is important to note that the maxim Leopold advocated here is inconsistent with the theorization of the land ethic. After enlarging the community concept to encompass all components of ecosystems, such as soils and waters, the maxim of the land ethic should be "to preserve the integrity, stability, and beauty of *the land community*." Compared with the biotic community, the land community is larger since it includes biotic communities and abiotic communities. It is unclear why Leopold adopted a narrower community concept to establish the maxim of the land ethic. However, according to the theoretical structure of the land ethic and the enlargement of the community concept Leopold repeatedly emphasized, it is justifiable to replace "the biotic community" with "the land community."

2 As for a detailed review of the development of the community concept in ecology, see Whittaker (1962), pp. 1-239.

3 See Gleason (1926), pp.7-26 and Gleason (1939), pp. 92-110.

4 A detailed discussion of Gleason's individualism on ecology can refer to McIntosh (1958), pp. 115-20, McIntosh (1967), pp. 130-87, McIntosh, (1975), p. 253-73, and Nicolson (1990), pp. 91-161.

5 Pickett and Ostfeld (1995), and Fiedler, White, and Liedy (1997), pp. 83-92.

6 See Moravec (1989), pp. 29-39.

7 Gleason (1939), pp. 107-08.

8 For instance, John Cairns defines integrity as "the maintenance of the community structure and function characteristic of a particular locale or deemed satisfactory to society." See Cairns (1975), p. 171 And James R. Karr avers that the concept of integrity implies "an unimpaired condition or the quality or state of being complete or undivided." See Karr (1997), p. 283. The difference between these two definition of integrity, as McIntosh notes, is that "[Cairns' definition is] a substantially anthropocentric meaning. By contrast, [Karr's definition is] with little or no influence from human actions." See McIntosh (2002), p. 66.

9 Gleason (1939), p. 99, emphasis added.

10 V. Grimm and C. Wissel identify 163 definitions of 70 stability concepts, and boil down these 163 definitions to 6 classes. They are: (1) constancy, (2) resilience, (3) persistence, (4) resistance, (5) elasticity, and (6) domain of attraction. See Grimm and Wissel (1997), pp. 324-26.

11 McIntosh (1985), p. 186.

12 McIntosh (2002), p. 72.

13 Gleason (1939), p. 97.

14 Ibid., p. 95.

15 Ibid.

16 Kirkman (2002), p. 76.

17 Ibid., p. 74.

18 Cafaro (2004), p. 101.

19 Leopold (1949), p. 203, pp. 224-25.

20 Norton has a general explanation of the relationship between uncertainty and sciences. He says: "On close look the 'problem' of uncertainty is really a grad bay of more or less related problems, all resulting from the fact that our finite knowledge will always fall short of any ideal of 'full' knowledge upon which the base everyday decisions. Uncertainty, in this sense, is just a general label for all the failures of our scientific models." See Norton (2005), p. 101.

21 Callicott (1999), p. 125.

22 Leopold (1949), p. 210.

23 Callicott (1999), pp. 123-24.

24 Leopold (1949), p. 203, emphasis added.

25 Ibid.

26 Ibid., p.205.

27 Ibid., p. 203.

28 Ibid.

29 Leopold (1979), p. 140.

30 Callicott (1989), pp. 88-89; and Flader (1974), pp. 143-44.

31 Leopold (1949), p. 173.

32 Ibid., p. 174.

33 Ibid., p. 223.

34 Ibid., p. 224.

35 Worster (1994), p. 429.

36 Callicott (1999), pp. 132-33.

37 Leopold (1949), p. 225.

38 McIntosh (2002), p. 66.

39 Shrader-Frechette and McCoy (1993), p. 37.

40 Callicott (1999), p. 125.

41 Ibid., p. 138.

42 Ibid., p. 136.

43 Ibid.

44 Ibid.

45 Leopold (1949), p. 217.

46 Ibid., p. 216, emphasis added.

47 Pickett and Ostfeld (1995), pp. 273-74, emphasis added.

48 Leopold said: "The circuit is not closed; some energy is dissipated in decay, some is added by absorption from the air, some is stored in soils, peats, and long-lived forests; but it is a

sustained circuit, like a slowly augmented revolving fund of life."
See Leopold (1949), p. 216.

49 Ibid., p.221.

50 Ibid., p.218.

51 Ibid., p. 220.

CHAPTER FIVE

Environmental Ethics and Bioethics: Anthropocentrism, Ideological Convergence, and Socio-Political Disposition

Edmund U. H. Sim

Introduction

The age-old issue of human relationship with the environment (and all living constituents therein) has in the past and at present remained a daunting challenge to mankind. This conundrum could be largely caused by a constantly changing, hence, arbitrary standards of ethical concerns for the environment and non-human sentient entities. Inadvertently, the fields of environmental ethics and bioethics face the affliction of rapid evolution in philosophical contemplation. Both are derivatives of environmental philosophy that have gained increasing prominence since the 1970s, possibly due to heighten academic and application interests. Major theories and philosophical approaches hitherto focus largely on the values (or worth) and rights-of-life of the natural environment. These are mostly interpreted based on worth of non-human living entities towards human (anthropocentrism), intrinsic value of non-human entities (irrespective of human-centered values, or non-anthropocentrism), or a loose merger of the two arguments. The first part of this paper argues for the deep (or obstinate) anthropocentrism in environmental ethics and bioethics. This leads to the second part which describes the growingly intertwined philosophical interpretation of environmental ethics and bioethics. Arguments are also introduced on how technological advancements in the fields of Biotechnology and Genetic Engineering have instigated the philosophical convergence of environmental ethics and bioethics. The third part of this paper touches on an emerging trend of argument that uses environmental ethics as a key point in global politics. It discusses how this phenomenon is viewed and used in political power-play among nations. Finally, from the information presented and debates prompted, it is hoped that the intellectual community will begin to academically tackle the emergence of new philosophies or the evolution of existing philosophies in ethics concerning the regulation of continuity and quality of all life on Earth.

Environmental Ethics and Bioethics Explained

The broad understanding of environmental ethics falls along the notion that the natural environment (with or without the presence of human) deserve due coverage in terms of ethical concern. In defining Environmental Ethics, Taylor (1989) explained that ethical consideration by human for the natural world is based on a moral concern for the environment. This is translated as our duties, obligations and responsibilities towards the natural environment including all flora and fauna inhabitants. In fact, the philosophy of moral relations between humans and the natural world is not a deliberate intellectual enterprise, but a concomitant response to nature in mankind's encounter with the environment (Rolston, 1988). Rolston (1988) exerts that recent philosophical argument on humans' responsibilities for nature is greatly influenced by the consequence of the scientific revolution that has allowed mankind to alter the development of natural history. Hence, the suitability of human towards Earth must entail a moral obligation to safeguard the continuity of the natural communities. This belief is most appropriately reasoned by Rolston in the statement,

> "......human inhabit natural communities as surely as they do cultural communities, and a major unfinished agenda in ethics is our responsibility for nature." (Rolston, 1988. p xii)

To further underscore its importance, Rolston portrayed environmental ethics as a revolutionary intellectual exploration that is fully and rapidly translated from theoretical platform to practical application. This is not normally evident for other philosophies or knowledge disciplines, especially with regards to making great impact in the world scene. As such, it should come as no surprise that environmental ethics would incur political interference – the essence of the third part of this paper.

Nevertheless, before embarking on a philosophical debate, it is necessary to take note of the theoretical development in the area of Environmental Ethics. Based on their review of Marshall's categorization of ethical approaches to the natural environment, Vardy and Grosch (1999) explained that academic contention on this matter rests on three main core beliefs. The first belief is the assumption that rights-of-life is not confined to human beings and should be extended to all inhabitants of the natural community. In the second belief, an interconnected and intertwined co-existence among all sentient entities is fundamentally necessary, and this ecological perspective depends on the availability of natural diversities. The third belief falls on the argument that ethical concerns for the natural environment are based on the usefulness of non-human biological entities to mankind. Hence, the moral obligation to safeguard the continued viability of biological diversities (conservation of the natural ecosystem) is vital to the sustainability of human life now and in future. This last belief is the most popular principle among all philosophical considerations, and represents the vanguard in the design of scientific, social, economical, and political agendas for environmental concerns today.

Incidental to the philosophy of environmental ethics is another increasingly prominent branch of study in ethics, commonly referred to as bioethics. It is selectively aligned to the disciplines of biology and medicine. The concept was first introduced by Fritz Jahr as early as 1927, and became a branch of learning, coined as Bioethics by Van Rensselaer Potter in 1970 (reviewed in Goldim, 2009). This ethical philosophy consolidates the practice of biological and medical research, and bio-engineering (biotechnology) applications with moral obligations and values. The early contemplation of bioethics is largely influenced by biomedical ethics that cover areas in human health science ranging from abortion to xenotransplantation. As development in the philosophy of bioethics progresses to include itself within environmental ethics, an effort was made by Potter to rename it as Global Bioethics in 1988 (Goldim, 2009). With this, bioethical concerns (according to Potter) can be redefined to take account of interaction among biology, ecology, medicine and human values for the universal survival of human and non-human organisms.

Obstinate Anthropocentrism

Ethical strategies (for the natural environment) that are drawn along the lines of human-centered premise gave rise to a philosophical viewpoint in environmental ethics that is viewed by some as "speciesism" and by most as anthropocentric ethics. In Vardy and Grosch's (1999) explanation, anthropocentric environmental ethics is divided into two perspectives, namely, strong and weak anthropocentrism. The former perspective argues that the human entity have to assume center stage in the reality of the natural system, while the latter contends that the best means of interpreting natural reality should emanate from a human-centered standpoint. Further to this, Norton argued that only weak (or extended) anthropocentrism permits humans to properly appreciate and obtain meaningful values from non-human biological entities in nature (reviewed in Afeissa, 2008).

It is not always true that philosophical contemplation of environmental ethics only revolve around anthropocentrism. Brennan (1984) advocated the ideology that as long as a natural entity (sentient and non-sentient) exists, it should be valued in any ethical or moral consideration. In explaining the need for environmental ethics, Brennan use the assertion of Jeremy Bentham on the concept of pain in non-human animals, and Peter Dobra's argument for higher order intelligence and social behaviors in cetaceans, as amongst the reasons necessitating humans to act as moral agent for non-human biological entities (Brennan, 1995). Commonalities between human and non-human living beings (irrespective of socio-economic values) form the basis of rationalizing environmental ethics. On the surface, such argument may appear ecocentric rather than anthropocentric. At closer examination, interpreting and defining intrinsic values (of non-human biotic entities) via the sensorial capabilities of human, reflects human-centeredness. For example, the delineation of intelligence hitherto is based on human-based logics. As long as some animals conform to similar structure of logical reasoning, they can be deemed as having a close resemblance to human. This then adequately justify their worth. Ironically, through this reasoning, amidst an ecocentric overtone, anthropocentrism is discernible. In addition, the behaviors and physiologies of many vertebrate members of the animal kingdom are closer to humans than organisms from the plant kingdom. Does this mean that the botanical constituents of the natural ecosystem are less relevant with respect to ethical worth? In fact, mankind today has great concerns for protection and conservation of plant resources in spite of the vastly dissimilar physiology and structure between plants and humans. It is obvious that ethical worth of natural resources (plants and animals) is often described as a function of anthropocentric agenda rather than the intrinsic values of non-human biotic entities.

Despite the many and varied arguments, the trend today for interpreting environmental ethics tends towards human-centered concerns. Conservationism (when envisaged) will emphasize anthropocentrism (strong or weak). No doubt, some philosophers remain adamant on the fundamental importance and ethical worth of intrinsic values in non-human biological beings. To them, such values need not align with usefulness towards humans (or values appreciated by humans), and the rights-to-life of non-human biotic entities can be justified on that ground alone (O'Neill, 1992). However, as pragmatism engulfs environmental ethics, anthropocentric proclivity will not be so easily excluded from the philosophy of ethics for the natural ecosystem.

Anthropocentrism in global bioethics runs deep and wide too. Although global bioethics has been distinguished from biomedical ethics to include ecology rather than just human biology and medicine (reviewed in Goldim, 2009), its manifested policies are not necessary ecocentric or nature-centered. Even if global bioethics can be partially ecocentric, antagonist of anthropocentrism must prove the existence of bioethics approaches that demonstrate tangible concerns towards the ecology (or non-human biotic entities) at the expense of human-centered values. In my opinion, such proofs are non-existence because it is inconceivable that humans are capable of understanding the emotion and psychology of animals and plants outside the logics and tools designed by humans. For example, it is an ethical requirement that informed consent must be solicited from human individuals who are chosen as subjects of biomedical experimentation. Such ethical practice is not enforced for animal testing in spite of a rigorous scrutiny on the study protocols by appropriate ethics committees. Humans decide whether other humans can conduct experiment on non-human animals/plants. The non-human biological entities do not have a say. Irrefutably, asking for consents from laboratory animals sounds of stupidity, but most (if not all) animals under testing do not willingly subject themselves for experimentation. Their struggle and trauma when restrained and probed by scientists are clear signs of non-consent, of which are often not heeded as such by humans. Even when bioethical guidelines (in research experimentation and application) on conduct towards non-human sentient beings may not associate with economic and health-related worth, the humane treatment of animals and plants feed the moral conscience of humans. Moral values are still a part of anthropocentric-based ethics. As long as humans lack the capacity to comprehend the intrinsic needs and reasoning of non-human biological entities, it is impossible that their (humans') policies on ethics can be non-anthropocentric.

Anthropocentrism is not a new thought process for defining environmental ethics and global bioethics, and is in fact often construed as a traditional western thinking influence. Evidences to this are provided by the accounts of early Greek philosopher, such as Aristotle, who maintained that the products of nature are created specifically for the interest of man; the contention of Thomas Aquinas that non-human animals are for the utility of mankind; and the thesis of the historian, Lynn White's on the role of Judeo-Christianity religion as instrumental in forging anthropocentric perspective on ethical concerns of the environment (Brennan and Lo, 2008). Eastern philosophies and thinking do not depart greatly from their western counterpart when it comes to environmental concerns and ethics. While often appearing as nature-centered and ecocentric at the outset, the appreciation of the environment and the preservation of non-human biotic entities (based on the eastern way of thinking), find ultimate motives in anthropocentric ends. For instance, the Hinduism and Buddhism teachings of perfect harmony with nature, and inclination towards optimal symbiosis with the ecosystem serve as means to upgrade afterlife or rebirth status of devotees. Most eastern way of thinking does not seek to preserve and venerate nature without tagging on to human-centered values.

Quintessentially, environmental ethics and global bioethics are fundamentally anthropocentric. It is this unifying factor that will be the primary impetus for the convergence of philosophies underlying the two fields.

Convergence of Environmental Ethics and Bioethics

It is traditionally conceived that despite their inter-relatedness, environmental ethics and bioethics are often addressed separately in academic and social discourse. In preceding arguments, I have established the cohesion of intellectual views between these two sub-disciplines of ethics through the concept of anthropocentrism. This unifying factor shapes their underlying philosophies which upon core analysis would reveal convergence. First of all, environmental ethics is devoted to ensure the perpetual viability of all sentient beings in order that ecological balance on Earth is ascertained. This balance is interpreted according to humans' evaluation of the natural order regulating the ecosystem. Secondly, while the ramification of global bioethics may not target the ecosystem in its original intent, it seeks to control the use of all sentient beings by mankind through a morally balanced approach. Such balance is judged according to human-based values and emotions. Finally, by taking these thrusts together, environmental ethics and global bioethics converged to sustain nature so that its preservation and utility meet the ecological and moral satisfaction of humans. In short, both disciplines use the philosophy of ethics on nature to serve anthropocentric agendas.

Different to anthropocentrism, but of rising concern, is the ecological perturbation on the natural environment due to technological advancements in biotechnology and genetic engineering. This is another fact that bridges environmental ethics with bioethics. The initial sphere of thinking for bioethics pertains to morally justified conducts by man towards himself and all other sentient beings in the affairs of biological and biomedical research (and applications thereof). It is mainly directed towards the humane treatment of all sentient entities. However, as the science of Genetics advances and the technologies of recombining genes improve, research endeavors yield new individuals (especially of animal and plant origins) that do not naturally exist in the normal environment. Many of these genetically modified creations are endowed with superior genes empowering them pre-eminence over naturally-occurring organisms in the competition for survival. When such phenomena appear in an ecosystem, ethics must be formulated to safeguard the natural integrity of the environment. The ethical considerations here no longer concern only bioethics but involve environmental ethics. Hence, instead of monitoring our utilization of the natural resources (environmental ethics) or our conduct in the manipulation of living things (bioethics), the regulation of what we introduced into the ecosystem becomes important too. Potter's redefinition of bioethics to global bioethics (reviewed in Goldim, 2009) is perhaps an earlier philosophical solution for the convergence of environmental ethics and bioethics. So, although it was never explained (by Potter) using similar reasoning, the intertwining philosophies between the two fields is not an unprecedented suspicion.

The infiltration of anthropocentric, ecological, and technological elements into philosophies underlying ethics for the natural ecosystem and all living things concomitantly prompted environmental ethics and bioethics to identify closely with environmental politics. In fact, since the future of environmental policies and bioethical concerns is strongly link to human society and livelihood, it is unsurprising that environmental ethics and bioethics will be subjected to socio-political intrusion. This is the focus of the next section. For the purpose of this paper, since the convergence of environmental ethics and global bioethics has been justified, it suffices to refer to both disciplines as just environmental ethics in subsequent discussion.

Socio-political Agenda

Politically-related philosophical movements in environmental ethics have been described in Brennan and Lo (2008). Accordingly, the rise in environmental politics during the 1980s in Europe entails a division in the major arguments into two groups – one proposing the "realists" agenda, and the other contending for the "fundamentalists" viewpoint. The "realists" is explained as practicing reformed environmentalism, having the standpoint of collaborative efforts among environmentalists, industries and governments towards controlled utilization of the environment. The ethics that prevailed would be translated as the mitigation of environmental pollution impact, the preservation of endangered species, and sustainable infrastructural development. This "soft" approach stands in stark contrast to the "fundamentalists" argument, where radical change is deemed as necessary to avert environmental devastation. Manifesting regulation should then take on priorities that are anti-capitalism and fanatical conservatism. While the "realists" perspective echoes a belief that anthropogenic environmental intervention is inevitable and that sustainable utilization is more logical than total conservation, "fundamentalists" argued against a subjugation of ethics and ideologies (on environmental protection) towards anthropocentric demands. These opposing arguments is perhaps a pragmatic reiteration of the concept of "shallow" and "deep" ecology movements introduced by the Norwegian philosopher, Arne Næss in the early 1970s. The "shallow ecology movement" fights for the effective reduction of environmental pollution and the obligatory control of natural resource depletion (Næss, 1973). In contrast, the "deep ecology movement", according to Næss, connotes a reverence for the intrinsic value of all natural entities independent of their importance to one another – a sort of a "biospheric egalitarianism".

The combined knowledge of the conceptual models that range from realism to deep ecology outlines an essential but peripheral premise for environmental philosophy in the context of socio-political discernment. A deeper contemplation on pragmatic interpretation of environmental politic theories must extend beyond policies of sustainable utilization or ideals of total conservation. It must adopt the bold approach of discussing environmental concerns (and ethics therein) within the context of tactical interplay in world politics.

With some third world countries rising to the status of developing nations, and numerous developing nations ascending to the level of developed or industrialized countries by the dawn of the 21st Century, the rapid progress of human civilization competes adversely with environmental preservation. The real awareness and burden-of-resolve to this is seemingly shouldered by developed and advanced nations, with the exception of a few. It is usually assumed that the endorsement for cooperative global resolution strategies adopted by developing and third world countries is more of international diplomacy gestures than genuine environmental concerns. Paradoxically, of the 190 countries (including the European Union) that have hitherto signed and ratified the Kyoto Protocol, 39 ($\approx$ 20%) are Annex I and II Party countries (developed nations and those with economy in transition) [Kyoto Protocol Status of Ratification (UNFCCC), 2009]. This does not include some recently industrialized countries such as Singapore, South Korea and Taiwan; and countries that now have high impact on global economy such as Brazil, China and India. The United States of America (USA) has signed to be included in the Kyoto Protocol, although it has yet to confirm ratification status. Taiwan has not taken any position in this matter. Adding these countries (excluding Taiwan) to the list of advanced nations that have committed to the Kyoto Protocol raised the proportion to around 44%. By this, actual developing and third world countries comprise 56% of those truly committed to the Kyoto Protocol. This is factual evidence that poorer nations or those with developing economies (mostly agro-based) are bigger players in the resolution of global warming compared to their richer and more industrialized counterparts. The concern here is that advanced and industrialized countries (and those with established economies) constitute no less than 78% of world total carbon dioxide emissions from the consumption of energy recorded in 2008. This information is processed from international energy statistics data made available by the U.S. Energy Information Administration (U.S. EIA, 2008).

Realism in this context has to be construed as the use of environmental ethics and policies for political bargaining and strategy. To elaborate, poorer nations lack the economic certainty and technological advantage to significantly mitigate environmental dilapidation due to anthropogenic climate disturbances. "Green technologies" and eco-friendly systems of harnessing natural energy are readily available to rich and advanced nations for reprieving the effects of global warming. Yet, socio-political commitments (on climate and environmental restoration) from richer and industrialized countries are not as forthcoming as from poorer and developing regions of the world. Antagonists (from developed countries) of the Kyoto Protocol have even cited injustice in blaming advanced countries; misguided premise inherent in the policy initiatives influencing the protocol; and the erroneous methods (designed for the protocol) to achieve effective results (Prins and Rayner, 2007). Irrespective of argued rationale, protagonist and antagonist must admit that the Kyoto Protocol represents symbolic expression of commitment by governments. The willing compliance of poorer nations with international commitment, sometimes initially, is at the expense of economic and infrastructural shortcomings. Nevertheless, it may be more important for these countries (or regions) to perceive their gain in political leverage (in the world stage) by exhibiting a higher ethical profile. Rich and technologically established countries may view global pledges as a liability on their political supremacy, especially when such commitments entail financial obligation. Two important valid assumptions have to be included for substantiating this line of argument. First, as the era of geographical territorialism has swiftly diminished due to globalization, the domination of foreign lands as a show of political prowess and technological might has become irrelevant. Second, the liberalization of political ideologies in many parts of the world coupled with the waning control (or absence) of colonies by world superpowers towards the end of the 20th Century, renders conflict among nations for supremacy in political ideologies an obsolete facet. The appetite of mankind for political dominance or recognition of such power is, nonetheless, insatiable. Therefore, the new reality on global competition and rationalization of national pride will fall on the weaponry of universal ethics (in our case, environmental ethics) in the battlefield of global altruism. Although the opportunity to compel advanced nations towards the table of negotiation (on environmental ethics) is perceived as an electoral advantage by the governments of some developing regions, it is a political liability for governments of advanced countries. Especially now, when the longstanding tradition of developed nations' advocacy and imposition of ethical values and policies (through environmental activism) on third world and developing regions is being challenged as duplicitous, and can be reversed to effect advanced countries. The recent United Nations' climate summit in Copenhagen [officially known as Conference of the Parties (COP) -15] led to an accord that carries a non-binding agreement with non-committal goals to reach new consensus of post-Kyoto treaty in reduction of greenhouse gases (Vidal, Stratton and Goldenberg, 2009; Papanicolaou and Fendick, 2010). This provides further proof of socio-political rivalry and tactical strategizing in global politics on the core issue of environmental ethics.

Realigning to formal theories in environmental ethics, I maintain that fundamentalism and "deep ecology" are most likely whimsical ideas that have no recourse on practical ethics. Realism and "shallow ecology" must engender pragmatism, of which, can only be completed by the inclusion of political strategy factors or "tacticalism".

Conclusion

Indispensable anthropocentrism, converging philosophies, and socio-political innateness seek rethinking of the concepts underlying environmental ethics and bioethics. In the bold effort to redefine the philosophy of these disciplines, the embodiment of realism and pragmatism must be explained in the context of obstinate anthropocentrism, ideological convergence, and socio-political "tacticalism". Inevitably and ultimately, the current theories to elucidate environmental ethics and bioethics cannot be assumed as immutable, and must be carefully dissected and rigorously re-argued.

Bibliography

Afeissa, H. S. (2008). The transformative value of ecological pragmatism. An introduction to the work of Bryan G. Norton. *S.A.P.I.E.N.S 1.1* (Available online, URL: http://sapiens.revues.org/index88.html; retrieved on April, 1 2010)

Brennan, A. (1984). The moral standing of natural objects. *Environmental Ethics 6*, 35-56.

Brennan, A. (1995). Ethics, ecology and economics. *Biodiversity and Conservation 4*, 798-811.

Brennan, A., & Lo, Y. S. (2008). Environmental ethics (Stanford Encyclopedia of Philosophy). (Available online, URL: http://plato.stanford.edu/entries/ethics-environmental/; retrieved on March, 25 2010).

Goldim, J. R. (2009). Revisiting the beginning of bioethics: The contribution of Fritz Jahr (1927). *Perspective in Biology and Medicine 52*(3), 377-380.

Næss, A., (1973). The shallow and the deep, long-range ecology movement. *Inquiry 16* (reprinted in 1995), 151-155.

O'Neill, J. (1992). The varieties of intrinsic value. *Monist 75*, 119-137.

Papanicolaou, C., & Fendick, L. (2010). Copenhagen summit fails to produce new global climate change treaty. *Lexology* (online). (Available online, URL: http://www.lexology.com/library/detail.aspx?g=456748b1-18cc047c9-a3c771004b1f0f; retrieved on March, 31 2010).

Prins, G., & Rayner, S. (2007). Time to ditch Kyoto. *Nature 449*(25), 973-975.

Rolston, H. III. (Ed.). (1988). *Ethics. Duties to and values in the natural world*. Temple University Press, Philadephia, USA.

Taylor, P. W. (Ed.). (1989). *Respect for nature. A theory of environmental ethics*. Princeton University Press, New Jersey, USA.

United Nations Framework Convention on Climate Change (UNFCCC). Kyoto Protocol Status of Ratification (December, 3 2009). (Available online, URL: http://unfccc.int/kyoto_protocol/status_of_ratification/items/2613txt.php; retrieved on March, 30 2010).

U.S. Energy Information Administration (U.S. EIA). International Energy Statistics (2008). (Available online, URL: http://tonto.eia.doe.gov/cfapps/ipdbproject/iedindex3/cfm?tid=90&pid=44&aid=8&cid..; retrieved on March, 30 2010).

Vardy, P., & Grosch, P. (Eds.) *The puzzle of ethics* (Rev. ed.) Fount Paperbacks, London, UK.

Vidal, J., Stratton, A., & Goldenberg, S. (December, 19 2009). Low targets, goals dropped: Copenhagen ends in failure. *Guardian* (online). (Available online, URL: http://www.guardian.co.uk/environment/2009/dec/18/copenhagen-deal; retrieved on March, 31 2010).

CHAPTER SIX

Sustainable Development vs. Sustainable Biosphere

Holmes Rolston, III

The United Nations Conference on Environment and Development entwined its twin concerns into "sustainable development." No one wants unsustainable development, and sustainable development has for the decade and a half since Rio remained the favored model. The duty seems unanimous, plain, and urgent. Only so can this good life continue. Over 150 nations have endorsed sustainable development.The World Business Council on Sustainable Development includes 130 of the world's largest corporations.

Proponents argue that sustainable development is useful just because it is a wide angle lens. The specifics of development are unspecified, giving peoples and nations the freedom and responsibility of self-development. This is an orienting concept that is at once directed and encompassing, a coalition-level policy that sets aspirations, thresholds, and allows pluralist strategies for their accomplishment.

Critics reply that sustainable development is just as likely to prove an umbrella concept that requires little but superficial agreement, bringing a constant illusion of consensus, glossing over deeper problems with a rhetorically engaging word. Seen at more depth, there are two poles, complements yet opposites. Economy can be prioritized, the usual case, and anything can be done to the environment, so long as the continuing development of the economy is not jeopardized thereby. The environment is kept in orbit with economics at the center.

One ought to develop (since that increases social welfare and the abundant life), and the environment will constrain that development if and only if a degrading environment might undermine ongoing development. The underlying conviction is that the trajectory of the industrial, technological, commercial world is generally right--only the developers in their enthusiasm have hitherto failed to recognize environmental constraints.

If economics is the driver, we will seek maximum harvests, using pesticides and herbicides on land, a bioindustrial model, pushing for bigger and more efficient agriculture, so long as this is sustainable. This will push to the limits the environmental constraints of dangerous pesticide and herbicide levels on land and in water, surface and ground water, favoring monocultures, typically of annuals, inviting soil erosion and invasive species. The model is extractive, commodification of the land. Land and resources are "natural capital."

At the other pole, the environment is prioritized. A "sustainable biosphere" model demands a baseline quality of environment. The economy must be worked out "within" such a policy for environmental quality objectives (clean air, water, stable agricultural soils, attractive residential landscapes, forests, mountains, rivers, rural lands, parks, wildlands, wildlife, renewable resources). Winds blow, rains fall, rivers flow, the sun shines, photosynthesis takes place, carbon recycles all over the landscape. These process have to be sustained. The economy must be kept within an environmental orbit. One ought to conserve nature, the ground-matrix of life. Development is desired, but even more, society must learn to live within the carrying capacity of its landscapes. The model is land as community.

"Sustainable" is an economic but also an environmental term. The Ecological Society of America advocates research and policy that will result in a "sustainable biosphere." "Achieving a sustainable biosphere is the single most important task facing humankind today" (Risser, Lubchenco, Levin, 1991). The fundamental flaw in "sustainable development" is that it sees the Earth as resource only. The underlying conviction in the sustainable biosphere model is that the current trajectory of the industrial, technological, commercial world is generally wrong, because it will inevitably overshoot. The environment is not some undesirable, unavoidable set of constraints. Rather, nature is the matrix of multiple values; many, even most of them are not counted in economic transactions. In a more inclusive accounting of what we wish to sustain, nature provides numerous other values (aesthetic experiences, biodiversity, sense of place and perspective), and these are getting left out. The <u>Millennium Ecosystem Assessment</u> explores this in great detail.

A central problem with contemporary global development is that the rich grow richer and the poor poorer. Many fear that this is neither ethical nor sustainable.

> Global inequalities in income increased in the 20th century by orders of magnitude out of proportion to anything experienced before. The distance between the incomes of the richest and poorest country was about 3 to 1 in 1820, 35 to 1 in 1950, 44 to 1 in 1973, and 72 to 1 in 1992 (United Nations Development Programme (UNDP), 2000, p. 6).

> For most of the world's poorest countries the past decade has continued a disheartening trend: not only have they failed to reduce poverty, but they are falling further behind rich countries (United Nations Development Programme (UNDP), 2005, p. 36).

The distribution of wealth raises complex issues of merit, luck, justice, charity, natural resources, national boundaries, global commons. But by any standards this seems unjustly disproportionate. The inevitable result stresses people on their landscapes, forcing environmental degradation, more tragedy of the commons, with instability and collapse. The rich and powerful are equally ready to exploit nature and people.

Such issues come under another inclusive term, "environmental justice." Now the claim is that social justice is so linked with environmental conservation that a more fair distribution of the world's wealth is required for any sustainable conservation even of rural landscapes, much less of wildlife and wildlands. Environmental ethicists may be faulted for overlooking the poor (often of a different race, class, or sex) in their concern to save the elephants. The livelihood of such poor may be adversely affected by the elephants, who trash their crops. Or it may be adversely affected because the pollution dump is located on their already degraded landscapes--and not in the backyard (or even on the national landscapes) of the rich. They may be poor because they are living on degraded landscapes. They are likely to remain poor, even if developers arrive, because they will be too poorly paid to break out of their poverty.

If we have trouble enough making sustainable development equitable, then, so much the more to be feared is emphasis on the sustainable biosphere--some will argue. Now the argument takes a new turn. The poor are kept poor because their development is not only constrained by the wealthy rich but by the setting aside of biodiversity reserves, forest reserves, hunting and catching limits. The priority of economics is the priority of human welfare, and that includes the welfare of the poor.

"Human beings are at the centre of concerns..." So the <u>Rio Declaration</u> begins, formulated at the United Nations Conference on Environment and Development (UNCED), and signed by almost every nation on Earth. This document was once to be called the <u>Earth Charter,</u> but the developing nations were more interested in asserting their rights to develop, more ecojustice, more aid from the North to the South, and only secondarily in saving the Earth. The Rio claim is, in many respects, quite true. The human species is causing all the concern. Environmental problems are people problems, not gorilla or sequoia problems. The problem is to get people into "a healthy and productive life in harmony with nature" (UNCED, 1992b).

Wilfred Beckerman and Joanna Pasek put it this way:

> The most important bequest we can make to posterity is to bequeath a decent society characterized by greater respect for human rights than is the case today. Furthermore, while this by no means excludes a concern for environmental developments--particularly those that many

people believe might seriously threaten future living standards--policies to deal with these developments must never be at the expense of the poorest people alive today. One could not be proud of policies that may preserve the environment for future generations if the costs of doing so are borne mainly by the poorest members of the present generation (Beckerman and Pasek, 2001, p. vi).

That is certainly humane, and no one wishes to argue that the poorest should bear the highest of these costs, while the rich gain the benefits. We are not proud of a conservation ethic that says: the rich should win, the poor lose.

But look at how this plays out with World Health Organization policy:

Priority given to human health raises an ethical dilemma if "health for all" conflicts with protecting the environment. ... Priority to ensuring human survival is taken as a first-order principle. Respect for nature and control of environmental degradation is a second-order principle, which must be observed unless it conflicts with the first-order principle of meeting survival needs (World Health Organization, Commission on Health and Environment, 1992, p. 4).

Again, that seems quite humane. But in India this policy certainly means no tigers. In Africa it means no rhinos. Both will only remain in Western zoos. To _preserve_, even to _conserve_, is going to mean to _reserve_. If there are biodiversity reserves, with humans on site or nearby, humans must limit their activities. Else there will always be some hungry persons, who would diminish the reserve. The continued existence in the wild of most of Earth's charismatic endangered species depends on some 600 major reserves for wildlife in some 80 countries (Riley and Riley, 2005). If these are not policed, the animals will not be there.

No. Keep some pocket reserves. Use them for eco-tourism, and the poor can benefit from the wildlife reserves on their lands. But the main driver is still going to be economics: sustainable development.

Economics is the overall governing driver; there is no escaping this, economists may say. For all of human history, we have been pushing back limits. Especially in the West, we have lived with a deep-seated belief that life will get better, that one should hope for abundance, and work toward obtaining it. Economists call such behavior "rational"; humans will maximize their capacity to exploit their resources. Moral persons will also maximize human satisfactions, at least those that support the good life, which must not just include food, clothing, and shelter, but an abundance, more and more goods and services that people want. Such growth is always desirable.

Some will say, if you wish to know what policy to sustain, you should ask an ecologist. Ecology is strikingly like medical science. Both are therapeutic sciences. Ecologists are responsible for environmental health, which is really another form of public health. Health is not just skin-in; it is skin-out too. One cannot be healthy in a sick environment. Health is something it is easy to advocate and the criteria seem to be scientific.

Any sustain-economic-development ethic, these ecologists may say, needs to be brought under a sustainable biosphere ethic. The fundamental concern is that any production of such goods be ecologically sustainable. Development concerns need to focus on natural support systems as much as they do people's needs. So "development," which has long been a concern and at which the West has been so successful in the modern epoch, is now entwined with, constrained by, "environment."

But ecologists have no special competence in evaluating whether to give priority to economic development or to conserving nature beyond what ecology is required for human development. A people on a landscape will have to make value judgments about how much original nature they have, or want, or wish to restore, and how much culturally modified nature they want, and whether it should be culturally modified this way or that. Ecologists may be able to tell us what our options are, what will work and what will not, what is the mininum baseline health of landscapes. But there is nothing in ecology <u>per se</u> that gives ecologists any authority or skills at making these further social decisions. Science does not enable us to choose between diverse options, all of which are scientifically possible.

I can equally substitute the word "economics" for "science" in what I have just been claiming. (Alternately put, "science" in the preceding claims, includes "economic science.") Economists have no special competence in evaluating what rebuilding of nature a culture desires, or how far the integrity of wild nature should be sacrificed to achieve this. Economists, like the ecologists, may be able to tell us what our options are, what will work and what will not. But there is nothing in economics <u>per se</u> that gives economists any authority or skills at making these further social decisions. Economics does not enable us to choose between diverse options, all of which are economically possible.

After four centuries during which economics has progressively illuminated us about how we can transform nature into the goods we want, the value questions raised in economics too are as sharp and as painful as ever. Economics can, and often does, serve noble interests. Economics can, and often does, become self-serving, a means of perpetuating injustice, of violating human rights, of making war, of degrading the environment. Religion and ethics do ask about how to live justly. We need religious insights into human nature as well as into nature. But here too both religion and ethics will be required to enlarge the scope of their classical concerns. Now, as did the ecologists and economists, the ethicists equally have their problems: caring for persons versus caring for nature. In the West we have built development into our concept of human rights: a right to self-development, to self-realization. Today, such an egalitarian ethic scales everybody up and drives an unsustainable world. When everybody seeks their own good, there is escalating consumption. But equally, if one seeks justice and charity, when everybody seeks everybody else's good, there is, again, escalating consumption.

Humans are not well equipped to deal with the sorts of global level problems we now face. The classical institutions--family, village, tribe, nation, agriculture, industry, law, medicine, even school and church have shorter horizons. Far-off descendants and distant races do not have much "biological hold" on us. Across the era of human evolution, little in our behavior affected those remote from us in time or in space, and natural selection shaped only our conduct toward those closer. Global threats require us to act in massive concert of which we are incapable. If so, humans may bear within themselves the seeds of their own destruction. More bluntly, more scientifically put: our genes, once enabling our adaptive fit, will in the next millennium prove mal-adaptive and destroy us.

Is there any hope? Humans are attracted to appeals to a better life, to quality of life, and if environmental ethics can persuade large numbers of persons that an environment with biodiversity, with wildness is a better world in which to live than one without these, then some progress is possible--using an appeal to still more enlightened self-interest, or perhaps better: to a more inclusive and comprehensive concept of human welfare. That will get us clear air, water, soil conservation, national parks, some wildlife reserves and bird sanctuaries. Environmental ethics cannot succeed without this, nor is this simply pragmatic; it is quite true. This may be the most we can do at global scales, even national scales, with collective human interests.

We may prove able to work out some incentive structures. The European Union has transcended national interests with surprising consensus about environmental issues. Kofi Annan, Secretary General of the United Nations, praised the Montreal Protocol, with its five revisions, widely adopted (191 nations) and implemented as the most successful international agreement yet. All the developed nations, except the United States and Australia, have signed the Kyoto Protocol. The Convention on International Trade in Endangered Species of Wild Fauna and Flora (CITES) has been signed by one hundred and twelve nations. There are over one hundred and fifty international agreements (conventions, treaties, protocols, etc.), registered with the United Nations, that deal directly with environmental problems (United Nations Environment Programme, 1997; Rummel-Bulska and Osafo, 1991).

Sustainable development is impossible without a sustainable biosphere. Thirty percent of the Millennium Ecosystem Assessment Development Goals depend on access to clean water. A third of the people on the planet lack readily available safe drinking water. Consider the conclusion of some of its principal authors:

> We lack a robust theoretical basis for linking ecological diversity to ecosystem dynamics and, in turn, to ecosystem services underlying human well-being. ... The most catastrophic changes in ecosystem services identified in the MA (Millennium Assessment) involved nonlinear or abrupt shifts. We lack the ability to predict thresholds for such changes, whether or not such a change may be reversible, and how individuals and societies will respond. ... Relations between ecosystem services and human well-being are poorly understood. One gap relates to the consequences of changes in ecosystem services for poverty reduction. The poor are most dependent on ecosystem services and vulnerable to their degradation (Carpenter, et al, 2006).

Pushing development in ignorance of the resulting outcomes on the poor, risking abrupt shifts at unknown thresholds past which the poor suffer much more degraded environments only further escalates the rich getting richer at the expense of the poor. The moral imperative is to keep the ecosystem services needed for the poor, even more than those needed for the rich. Since ecosystem services are involved for persons living immediately in contact with nature, such conservation is as likely to focus on a sustainable biosphere as on sustainable development.

The fundamental flaw in sustainable development as first priority is moral in yet a deeper sense. Ecologists, economists--and ethicists and theologians--alike need to learn that there is something morally naive about living in a reference frame where one species takes itself as absolute and values everything else relative to its utility, even if we phrase it that we are taking ourselves, rich or poor, as primary and everything else as secondary. If true to their specific epithet, ought not Homo sapiens value this host of life as something with a claim to care in its own right? If we humans continue as we are headed and cause extinctions surpassing anything previously found on Earth, then future generations, rich or poor, are not likely to be proud of our destroying this wonderland planet we have been given.

Develop! Develop! Develop! Maximize endless development? The theme of this AAAS Convention is Our Planet and Its Life: Origins and Futures. Is the future we want maximized development for human satisfaction? Perhaps when we couple origins and futures, we will in the midst of our development also seek to sustain life on this wonderland planet. People and their Earth have entwined destinies; that past truth continues in the present, and will remain a pivotal concern in the new millennium.

References:

Beckerman, Wilfred, and Joanna Pasek, 2001. Justice, Posterity, and the Environment. New York: Oxford University Press.

Carpenter, Stephen R. et al, 2006. "Millennium Ecosystem Assessment: Research Needs," Science 314(13 October):257-258.

Riley, Laura and William Riley, 2005. Nature's Strongholds: The World's Great Wildlife Reserves. Princeton, NJ: Princeton University Press.

Risser, Paul G., Jane Lubchenco, and Samuel A. Levin, 1991. "Biological Research Priorities--A Sustainable Biosphere," BioScience 47:625-627.

Rummel-Bulska, Iwona and Seth Osafo, eds., 1991. Selected Multilateral Treaties in the Field of the Environment, II. Cambridge: Grotius Publications.

United Nations Conference on Environment and Development (UNCED), 1992b. The Rio Declaration. UNCED Document A/CONF.151/5/Rev. 1, 13 June.

United Nations Environment Programme, 1997. Register of International Treaties and Other Agreements in the Field of the Environment. Nairobi: United Nations Environment Programme.

United Nations Development Programme (UNDP), 2000. Human Development Report 2000. Oxford, UK: Oxford University Press.

United Nations Development Programme (UNDP), 2005. Human Development Report 2005. New York: United Nations Development Programme.

World Health Organization, Commission on Health and Environment, 1992. Our Planet, Our Health: Report of the WHO Commission on Health and Environment. Geneva: World Health Organization.

CHAPTER SEVEN

The Possibility of a Global Environmental Ethics:

A Confucian Proposal[1]

Shui Chuen Lee

Introduction

The fast development of communication technology, both the real and the virtual connections that we built around our lives, indicates that we are all living in a closely knitted global village. Anything happening in one corner of the world could reach anyone who wants to know and, no matter how insignificant it seems to world politics, its butterfly effect may loom large when it spreads through the internet. The development of genomic study reveals another aspect of our closeness between each other. In fact genetic studies tell us more than that as *Homo sapiens* we are genetically one whole species, but we are also very closely related with all other species on Earth. Our species identity and consequently our individual identity could not be understood without the background of our Earth and the biological circle on Earth. As a member of the human family, we are genetically, biologically and socially in one family all along. Diversity is only an outgrowth of our underlying identity.

Furthermore, the effect of our interactions and influences is no better exhibited by the experience we face today in our environmental crises. The recent climate warming is a threatening case. The outcome is a cumulating effect of more than two hundred years and over a very large stretch of land and countries. The disaster is felt by every nation and in fact upon everyone on Earth sooner or later. Such global disasters could not be solved by any individual country alone. We have no choice but to orchestrate our efforts and need to take up the point of view of a global citizen to deal and hope to save our common future. The old individualistic conception of human life and affairs, whether of the individual or individual nation, is not only incapable to deal with such problems, but is also a false conception of our reality since we are one whole ecological complex on Earth.

We have had a hot debate in the eighties of the twentieth century about the place and standing of human interest in environmental ethics. The West is dominated by liberal individualism as a whole, but we soon understand that environmental problems could not be solved by this model. For such a model would not post any responsibility for us to take care of the benefit of future people, nor the interest of members of other species.[2] However, since then, it has been successfully argued and I believe firmly established by environmental ethicists that anthropocentrism is wrong and we have to take a holistic view in our environmental issues. The fairly success of the Land ethic signifies at least that people concerning the environmental problems have more or less adopted a holistic world view. However, it seems that the most difficult task is to get people of different societies, different nations and cultures to work together. It seems that we still have to work out a global consensus across different regions and cultures to make good our common goal. In this paper, I try to propose a Confucian way of looking at our present day environmental problems and see how it could provide some new inkling to work on.

1. Some Basic Facts of Reason: Global Citizenship and Global Conditions

We have to accept that the environmental problems that we face today are inimical to our survival and to many of the species on Earth. To name a few, the destruction of rainforest, the shortage of fossil fuels, the exceptionally large human population exceeding the carrying power of the Earth and so on are grave issues of survival that we have to face today. We have to recognize that without proper response to these problems the result would be disastrous and human being would be one of the species that suffers the most. Furthermore, such global disasters could not be solved by individual nation or groups of people as pollution finds no border and must be solved by global cooperation without taking advantage of each other. Human beings have to accept that we are all in a family, that is, we are global citizens and each has to play out his or her duty to make good the whole environment.

This global citizenship must be extended to all living things. We are not only genetically closely knitted with all other species, we are also sharing a holistic ecological pyramid and very much inter-dependent. Our interest as well as our identity is bound up with all other species. There is mutual affinity between living things. We have certain kind of feeling of affinity with other species and often we are deeply touched by the sufferings of non-human animals. This feeling of empathy elicits human beings out of their self-centered way of thinking and urges them to judge with a trans-species point of view. Such trans-species commitment to the welfare of all does help to raise the moral status of human being without inappropriate species discriminations. As the most intelligible and morally sensitive species on Earth, we are the self-conscious part of mother Earth and have to take responsibility for our own actions as well as the wellbeing of other species or the whole ecological complex on Earth. We are naturally moral agents. Though we may value our species most, we have to accept that other species have values independent of our judgment. Now, any species discrimination is anthropocentrism without reason, and as a rational being, we could not accept such actions ourselves. Hence, we have to renounce the deep underlying unconscious belief that we may sacrifice non-human lives for our own good. We have to take other species as equally worth our considerations. It is no longer reasonable to take our own interest as paramount and could not treat other species as merely instruments for our own benefits.

As a moral agent we have a unique role to play in this family of life. We are in a sense ranked as one of the highest beings of the ecological pyramid; hence, we have to take up much more responsibility towards the whole. It is only through a species unselfishness way of treating others that we may justifiably win the respect of other species without being charged as anthropocentricism or specieism[3]. We could not relegate our duty lightly when we could act either by refraining from certain actions or by promoting certain effort to the prevention of the extinguishing of other species. We bear the responsibility if any harm to these species including their being extinguished is due to human improper actions. As a species with moral capability, we could not but have to judge morally not only in human affairs but in environmental issues. We are in a sense bounded by a natural moral community with other species. There are some minimal moral rules in our dealings with ourselves and with other species.

The basic element of this re-conception is the recognition that other non-human species share our identity and our common interest. We have to cross not only the national border but also the species border to let other species form with us a kingdom of common interest. They are forming part of our kingdom of ends. They are in fact forming a holistic unity with us though loosely and sometimes with conflicts. Not only that our interest is bound up with theirs, theirs is also a major factor in our estimation of what we should act to the remedy of our present environmental crises. We are so to say, not only a steward of their existence, we are also the nurturer of all living things. For example, we show our deep concern with the interest of other species by conservation and by restoration of the environment and increasing the number of those endangered species. We assume the duty to maintain the healthy growth and subsistence not only our own species but also other species and sometimes need to act further to promote their goods with certain sacrifice of our own interests.

Last but not the least is the recognition that we have limited living resources and each species in the ecological pyramid has certain natural limitation that no one could be exempted. We have learnt by the alarming results of environmental deteriorations throughout the twentieth century to respect the natural order and natural development of the biosphere and that we should not disrupt the natural process unless we have good reasons to do so. The evolutionary process is a summary of the natural limit of species development on Earth. For instance, we have to rely on each other in the food chain of the pyramid. We have to be alerted that some of our ways of living or standards of living are not sustainable and need be changed for good. We have to cooperate as global citizens. To make our actions effective, we need to refrain from both asking too much or too less for our own and for others in our daily lives. We could not ask everyone, say, to refrain from eating meats. Nor could we accept letting people exploit the environment as much as it is within their possession. We have to be holistic in general and let individualistic preference in particular cases be coexisting as far as possible without toppling the whole ecological pyramid.

2. A Confucian Conception of Holistic Perspective: The Relation of Man and Nature

The three most important philosophies of Chinese tradition, namely the Confucianism, Taoism and Buddhism are without exception holistic and taking other non-human lives as having independent values by their own. Confucius once said,

> Heaven does not speak. The four seasons go and myriad things flourish. Heaven does not speak indeed. (*Analects* 17:19)[4]

Heaven or *tao* does not speak out but shows in the natural process the growing and flourishing of lives. This means that *tao* (Heaven) endorses everything with value, not just for human beings. Later in the *Doctrine of Means* it is said that the *tao* of man is to realize the *tao* of heaven and it means to sustain the growing and flourishing of all things (*Doctrine of Means*[5], chapter 22). The famous and representative figure of Song-Ming NeoConfucians, Cheng Ming-tao said explicitly that a man of *ren*, or a moral person, is empathetically united with everything under heaven and earth[6]. Confucianism regards human being as one with all living things under heaven or throughout the whole universe. We are both ontologically united with all species and practically one by all means. In the evolutionary process, it is indeed thanks to the preceding species that *Homo sapiens* could emerge as a specific kind of being. However, for Confucianism, human being plays a unique role in this ladder of existence because man is the only moral agent on Earth. This means that according to Confucianism we have to take up our moral responsibility towards the wellbeing of other species. The *Doctrine of Means* gives a clear picture of the relation and duty of man towards nature as follows:

> Only those who are absolutely sincere can fully develop their nature. If they can fully develop their nature, they can fully develop the nature of others. If they can fully develop the nature of others, they can fully develop the nature of things. If they can fully develop the nature of things, they can assist in the transforming and nourishing process of Heaven and Earth. If they can assist in the transforming and nourishing process of Heaven and Earth, they can thus form a trinity with Heaven and Earth. (*Doctrine of Means,* Chapter.22)

To act sincerely is to act authentically in accordance with our natural endowment of intelligence and moral capacity, that is, to act unselfishly as a moral agent. To fully develop our nature means to act according to *tao*, that is, to promote the wellbeing and flourishing of others and other species. For Confucianism, as a moral agent we have a moral mandate to live out our moral ideal and this is what makes human being morally respectful and have a life with dignity. This self imposed moral command is regarded by Confucianism as the most significant inborn human nature. The fulfillment of this moral or Heaven mandate transfers us above our natural species boundary and become united with *tao* or feeling oneness with the universe. Moral capacity endows us with dignity. It is the morally sincere person who is true to his or her inborn moral nature and who could fully fulfill this moral mandate. Since we are free, we could share and participate voluntarily the nourishing process of Heaven and Earth. It is because of this capability we have the responsibility to take care of everything on Earth and serve not only as a steward but also a nurturer like *tao* which nurtures and sustains everything. It implies that we have responsibility not only to our contemporaries but also to our posterity, to the flourishing of the future generations of *Homo sapiens* as well as other species. If we achieve such virtue we are the sage and united with Heaven and Earth in a holistic unity.

Philosophically, the basic moral concept of Confucianism is Confucius' conception of *ren* (humanity) or Mencius' conception of "the mind that could not bear the sufferings of others.(2A:6)[7]" Both of them signify our natural feeling of empathy with each other and with other species. It is our internal natural moral consciousness towards the sufferings of others. When we see or even just learn indirectly somebody or some animal badly hurt or cruelly treated, we could not but feel morally uneasy and disturbed in our mind. This moral consciousness is also the source of our moral command that prompts us to take action to relieve the pain or promote the wellbeing of others. For Confucianism, the way to carry out this self imposed moral mission is to act through *li* (moral principles and rituals) in a humanly and properly considered pattern. Hence, moral principles and rules are the channels that help to actualize our moral commands. The particular way to act according to principles, especially the specific form of ritual or action could be amended according to the need of the situation and the call of humanity (*Analects*, 2:23). It depends largely upon our moral experience and our grasp of the morally relevant elements in the context. It comes close to Aristotle's idea of practical wisdom though the guiding source is always our moral mind of *ren*. *Ren* is first expressed naturally in our early years as filial piety to our parents and respect to our elders (*Analects*, 1:2). It is reflected in our love of and passions with our family members. To them, we bear the most intimate relationship and the most primordial moral duties. However, it is only the starting place of our practice of morality. Our moral responsibility is not limited to intimate family members but also extended towards others. Mencius said,

> We have the duty to respect our elders first and then extend it to other's elder; to nurture our youth first and extend it to other's youth (*Mencius*, 1A:4).

Hence, we have a gradation or diffusion of duties to others. It is a principle of differentiation of the performance of our moral duties. We have more stringent duties towards our parents and children. However, we have to extend it to our fellow citizens, different peoples and other species. It would ultimately extend all the way to cover everything on Earth, nay, to the whole universe. It implies that we have to treat our fellow living things fairly as they are endowed with *tao* like us. With such a mind of *ren* or humanity, we have to fulfill as much as a moral agent have to, that is, to nurture as far as possible. Mencius captures this holistic responsibility in his further emphasis of the order of our duties relative to our family members, to others and to other species in the following:

> Be intimate with family members and going on to treat people with humanity; treating others with humanity and going on to be beneficial to all things (*Mencius*, 7A:46)

This is a broad guiding principle. Confucians are very much aware that the decision- making in practical affairs are not easy. To make the most prudent and moral judgment in the day to day affairs is the chief goal of moral cultivation. It is what we have to learn by doing. Thus, Confucianism is rather down to earth in personal and global decisions and would make truly workable solutions that reflect our authentic moral experience and our understanding of the objective reality.

In sum, Confucianism recognizes Heaven and Earth as the carrying mother, which creates and nurtures all things. We are one among the myriad things and have to respect the value of each living thing and each species. Though we could not overstep the natural circle of ecology, human being has to follow the step and spirit of mother Earth to nurture all things and thus serving the kingdom of nature as a nurturer. Confucius once remarks that it is human being that could enlarge *tao*, not that *tao* enlarges man (*Analects*,15:28). It means that human being is the one who could put *tao* into action and realize it on Earth. Human being is sometimes heralded as the heart of Heaven and Earth[8] to signify its special moral status as well as its moral responsibility to all species. Confucianism may be said to give a moral justification of the notion of ecological conscience in Land ethic.

3. How to Make Good of Ecological Pluralism: Harmony with Difference and Some Working Principles

For Confucians, there are a number of middle principles that are essential and very helpful to make our practice in environmental ethics work individually and holistically. The first one is the principle of extending the nature of everything to the utmost which is embedded in one of the passages of the *Doctrine of Means* quoted before. I would like to explore further its rich implications in our practical actions:

> Only those who are absolutely sincere can fully develop their nature. If they can fully develop their nature, they can fully develop the nature of others. If they can fully develop the nature of others, they can fully develop the nature of things. If they can fully develop the nature of things, they can assist in the transforming and nourishing process of Heaven and Earth. If they can assist in the transforming and nourishing process of Heaven and Earth, they can thus form a trinity with Heaven and Earth. (*The Doctrine of Mean*, Chapter 22)

This basic principle expresses the fair treatment of everything on Earth. The so-called "nature" means the endowed potentials and capabilities of each living thing. For human being it also manifests as our moral striving. This principle requires first that we let everybody have the chance to develop to the utmost his or her talents as well as let our moral sensitivity express freely without hampering by social or natural conditions. It contains also an idea of equality and equity that we have to respect other's freedom to develop their endowment as far as possible. In case of conflict, we have to make certain specification and balance between the contenders. Furthermore, as we have argued, embedded in this principle is our duty as the nurturer of other species as well.

Another important principle is the principle of harmony with difference. It develops from Confucius saying that we do not require everybody to do the same or attain the same end. We shall let everyone develop their own interests and ideals as far as possible with the conditions that we could live harmoniously together. It means that we should not take authoritarian means to solve our conflicts and let everybody and indeed everything retain their individual difference as far as possible. Thus the different traditions and preferences of each people and the natural ways of living of other species should be maintained as far as possible. In case of conflict, whether it is within our species or between species, Confucianism will try to solve it peacefully according to what is available and affordable at hand.

A third principle is the Confucian way of expressing our individual relationship with others. It is the principle that we may express our love and responsibility according to our different relationship with others. Each one should act with an eye upon the duties coming with their natural and social inter-personal relations. For example, our duties to our parents and children has a higher bearing than our friends and fellows, and our duties to the latter would be higher than those to other species and further still to the ecological elements. This differential gradation of duties and responsibilities is accorded fairly for everybody. It is the duty of each moral agent to make good of her or his case in particular situations without damage to others and respecting the interests and rights of others as far as possible.

In response to the shortage of resources for the satisfaction of everybody, Mencius proposes a fourth principle that we should lead a life without too much demand of satisfaction of our desire. This is a way to let our moral commands be carried out more easily. Besides, we shall have more spiritual enjoyment and less material distractions to our moral virtues. Reducing unnecessary desires and satisfactions is also important to maintain a sustainable social and natural environment. For instance, in face of the too large a human population, we have to make certain self restraint to reduce our numbers in the future so as to let our dear mother Earth retain its carrying power in time and our future generations, including other species as well, to have an enjoyable environment.

In general, we have to make allowance for our individual capabilities and our responsibilities to our fellow persons as well as other species. We have to take seriously the urgency of the global problems that we face together today and have to limit our interest within the moral bondage of a global citizen. The commons are for us as well as for other species, for us as well as for our future generations. Any selfish exploitation of the commons is unsustainable and against our rational and moral requirements. With this moral framework, Confucianism may provide the common ground for our collective effort and co-operation between nations and people to the preservation of our living environment and to the best of our common future.

4. Some Ramifications in the Global Warming Issue: Towards an Ethics for Global Environmental Problems

Environmental problems have been a major concern for scholars east and west for the last decades, and international efforts have been organized to solve these problems with at least limited success. For instance, the recognition of existence and harmful effect of the ooze hole and the forbidden of further production and use of CFC products is a case heralded as a success of international co-operation and provides a good example for future solution of similar global problems. The global efforts in restoration of endangered species and prevention of the destruction of rainforests are more or less recognized as our common responsibility. However, the underlying ground for such cooperation is mainly national interest and it is because of the obvious scientific certainty of the fault of CFC and the relative small scale of the backlash on economic development that developed countries were quick to come to a consensus banning its use. It may be fortunate that at that time we have some alternative products ready to make a substitute. However, when it comes to more serious national interest and diffusing cause and responsibility, such as the global warming problem, international co-operation is much more difficult. We have such an experience in the signing of Kyoto protocol and the carrying out of the Agenda 21. As many studies have shown that when the issue is not quite clear about its cause and effect, when the global community is quite fragmented in its evaluations, and when the threatening result is in the distant future, it would be very difficult to arrive at global and orchestrating actions. Since it is much less cared by individuals and individual countries and the wager seems heavy, the needed actions would not be easily agreed upon. When it is coupled with huge national economic interests, it is almost certainly doomed to fail. Global warming is just such a case in point. It is until the effect is too obvious and the damage to other peoples or nations, or most importantly to every nation, becomes a glaring reality, that more and more people throughout the world start to press for actions, that leaders of nations have the interest to face it and willing to make a deal.

It is now obvious that global warming has great destructive effects on the environment and upon our low sea level fellows and very many types of species. It will make our future generation much harder to survive and in a harder living conditions, such as shortage of clean water, not enough food to feed and less useable land to live. We are not only unable to live up to our mandate as a nurturer, we are much less than a steward of our Earth. We are in fact living against the moderate demand of sustainability. Since it is also obvious that the present global warming effect is at least partly if not wholly anthropogenic, we have to take up our responsibility. What human has done could be undone and should be undone if it is wrong. It is no excuse even what we could do may not be able to avoid the tragedy happening. It is just like saying that all species may ultimately not be able to escape the fate of warm death or that the Earth would vanish in the expansion of the sun in the distant future, but this does not suspend our moral responsibility a moment less. As a member of a holistic whole we have to admit our responsibility even though we seem to live innocently as an ordinary person. It could not be ducked that our individual contribution is too insignificant and has little responsibility towards such gigantic deterioration of others and other species. As a species of over six billion members, our collective action is as effective as any mass motion on Earth. Our individual action of redemption will not be in vain.

In fact, with our powerful technology and know-how, we could make a great change. The only obstacle is how to make up our determination for a global orchestrating action to solve this problem. In the past, the United States of America was no doubt the one most irresponsible and greatest obstacle for the global cooperation in its solution. The unilateralism of USA is the most obvious kind of national selfishness against the wellbeing of other peoples and other nations. The delay of the global rectification of the Kyoto Protocol leads global warming problem almost to a point of no return. The increase of green house gas, especially carbon dioxide, over the last few years, helps to fuel global warming and very difficult and much less successful to make good now.

The 2009 Copenhagen Accord gives a new hope for the realization of cooperation though with many controversies and conflicts of interests of different nations and parties. Somehow it makes a step forward. Further actions of the governments of nations on Earth need be observed for the next few months, before we could work out some collective and efficient actions to solve this mounting issue. Though international justice between the developed and underdeveloped countries need be contained, certain compromise by all is a necessary precondition for a really effective solution. The basic value is that the future generation is for us all. There should not be and will not be nationally or ethnically divided *Homo sapiens*. Our sacrifice is sacrifice for our future generation. We could do more if we take them as our family members. In fact, we are predicting there will be great fights in the coming age of shortage of materials for living. Wars are what we must avoid morally and rationally. There are a number of immediate actions that need be taken individually and globally. First of all, we need to reduce the carbon consumption by all means individually socially and internationally through low carbon way of living and new technologies. Second, it is our imperative to join hand in hand to reduce the world population in the near future. Third, we have to consider the release of national borders to admit those who are going to lose their homeland because of rising sea level and because of the destruction by violent climate changes. The abolition of national border is inevitable from a holistic point of view and the integration of all people is true to the spirit of the notion of a common future for all.

It would be a great lesson and success for all if this and other grave environmental problems lead us to reflect more deeply our intra-species and inter-species relationship, the socalled human-nature relationship[9]. Such grand issues have the power to wake us up from our ethnic centered, anthropocentric way of doing things. It would be a great achievement if we finally learn how to live peacefully and equitably together with our fellow people and fellow species. By this we may realize the long time dream of our fore-runners: the perpetual peace between nations and the harmonious utopia of equity and flourishing heaven for all.

Reference

Brundtland, Gro Harlem World Commission on Environment and Development. Brundtland Commission, *Our Common Future.* Oxford/New York: Oxford University Press, 1987.

Callicott, J. Baird, *Earth's Insight: A Survey of Ecological Ethics from the Mediterranean Basin to the Australian Outback.* Berkeley: University of California Press, 1994.

Chan, Wing-tsit, *A Source Book in Chinese Philosophy.* Princeton: Princeton University Press, 1963.

Leopold, Aldo, *A Sandy County Almanac and Sketches Here and There.* New York: Oxford University Press, 1949.

Naess, Arne " The Shallow and the Deep, Long-Range Ecological Movement," *Inquiry* 16 (Spring 1973).

Newton, Lisa H., *Ethics and Sustainability: Sustainable Development and the Moral Life.* Upper Saddle River, N.J.: Prentice Hall, 2003.

Regan, Tom, eds. *Earthbound: New Introductory Essays in Environmental Ethics.* New York: Random House, 1984.

Peter Singer, Animal Liberation. New York: Random House, 1975.

Tu, Wei-ming, *Centrality and Commonality: An Essay on Chung Yung.* Monograph of the Chinese and Comparative Philosophy, no. 3. Honolulu: University Press of Hawaii, 1976.

Tucker, Mary Evelyn, and Tu Wei-ming, eds. *Confucianism and Ecology: The Interrelation of Heaven, Earth, and Humans.* Cambridge, M.A.: Harvard University Press, 1998.

Endnotes

1. An earlier version of this paper had been presented in the "Applied Ethics: Third International Conference at Sapporo" on November, 21-23, 2008, organized by Hokkaido University, Sapporo, Japan.
2. Cf. Tom Regan edited *Earthbound: New Introductory Essays in Environmental Ethics* (New York: Random House, 1984), pp.3-37.
3. Cf. Peter Singer, *Animal Liberation* (New York: Random House, 1975).
4. This and later translations of Chinese classics are basically mine with reference to Wing-tsit Chan, *A Source Book in Chinese Philosophy* (Princeton: Princeton University Press, 1963).
5. It is one of the socalled Confucian classic "Four books" and is sometimes translated as *Chung Yung*. There is a full translation in *A Source Book in Chinese Philosophy*. Some of the exploration of the meaning of these and other sayings please referred to Wei-ming Tu, *Centrality and Commonality: An Essay on Chung Yung* (Honolulu: University Press of Hawaii, 1976).
6. It appears in one of the famous letter " On Understanding the Nature of *Jen*" wrote by Cheng Ming-tao, which is translated in full in *A Source Book in Chinese Philosophy*, pp.523-4.
7. *A Source Book in Chinese Philosophy*, p.65.
8. This phrase was coined by the famous Ming Confucian Wang Yang-Ming and is the common thread running through Confucianism from Confucius and Mencius up to contemporary NeoCofucians. Cf. *A Source Book in Chinese Philosophy*, p.690.
9. I have treated this subject in Confucian terms in an earlier paper, "The Fundamental Ideas of a Confucian Environmental Ethics: A Critical Response to Callicott's Project" in the Environmental Conference of " Heaven, Earth

and Human: The Trios of Environmental Ethics" on August 4, 2000 at Chungli, Taiwan, organized by the Graduate Institute of Philosophy, National Central University, Taiwan, R.O.C. A Chinese version of it was published in *The Legion Journal*, vol.25 (Taipei: Legion Monthly Publisher, December, 2000), pp.189-205.

CHAPTER EIGHT

Confucian Filial Piety and
Environmental Sustainability

A. T. Nuyen

China is developing at a break-neck speed. Its rate of economic growth has averaged about 10% annually in recent years and is looking set to continue at this rate for the next few years. Compounding at this rate, the economy will double in size in about seven years. Observers have pointed out that China's economic achievement has been at considerable environmental costs and some have predicted that environmental problems will limit economic growth in the future, arguing that economic development is sustainable only if it is environmentally sustainable. For now, it seems that environmental sustainability is not a major concern for the Chinese rulers. Critics point to the official Chinese position at the recent conference on climate change in Copenhagen, which is unsupportive of strong measures to tackle global warming, as evidence of this lack of concern. Given this background, it may seem wrong-headed to go into China's cultural heritage, namely Confucianism, in search of environmental insights. However, it is at least arguable that China has abandoned Confucian principles in its pursuit of economic growth and that the way to get the Chinese people and their government to set their sight on environmental sustainability is precisely to appeal to their cultural heritage, or to Confucianism. The task is urgent insofar as global sustainability will be unachievable without the participation of one fifth of the world's population. This claim, of course, is predicated on the assumption that Confucianism calls for a commitment to environmental sustainability. It is this assumption that I want to verify in this paper. Specifically, I will argue that the commitment to filial piety, a virtue strongly endorsed by Confucians, entails a commitment to environmental sustainability. I will discuss Confucian filial piety in Section I. In Section II, I will show how filial piety entails an obligation to adopt environmental practices that ensure environmental sustainability, following a procedure developed by Onora O'Neill. In Section III, I will examine some textual evidence that may be cited in support of my claim that Confucianism is committed to sustainable environmental practices.

I

In the *Analects*, filial piety (*xiao*) is construed as a duty primarily towards one's parents. However, it is seen also as a duty towards one's ancestors by extension. Towards one's parents, the duty is to be reverential, not just to take care of their needs: "Nowadays, *xiao* is taken to mean that one is able to provide for one's parents. Yet even dogs and horses receive as much care. Without reverence, what is the difference?" (*Analects* 2.7).[1] How one is to show reverence to one's parents is set out in the rites, or rituals (*li*): "When your parents are alive, serve them according to *li*; when they are dead, bury them and sacrifice to them according to *li*" (*Analects* 2.5). Naturally, "according to *li*" has to be understood as following rituals with a reverential attitude rather than simply following rituals. As Confucius said, "What can I find worthy ... in a man who is lacking ... in reverence when performing the rites...?" (*Analects* 3.26).

Filial piety is a crucial Confucian virtue even though commentators are divided over the question of whether it or some other virtue, such as benevolence (*ren*), is the most fundamental. When Mencius says that filial piety is the "actuality (*shi*) of benevolence" (*Mencius* 4A27),[2] he seems to think of it as the more important virtue. The most explicit endorsement of the view that filial piety is the foundational virtue can be found near the beginning of the *Analects:* "Filial piety and brotherly respect – are these not the root of perfect goodness!" (1.2). In the *Xiaojing* (*Classic of Filial Piety*), filial piety is seen as the "root" of all virtues. Relying on passages such as these, Ivanhoe takes the idea of being the "root" as an important feature of filial piety.[3] He identifies two other key features. The first of these two is that filial piety "is not just a general feeling of gratitude" towards one's parents for their kindness: it is "partially constituted by the sense that this kindness was done by someone who was dramatically more powerful than oneself and who sacrificed substantial goods of their own in order to care for one" (p.196). The final feature is "the intimate and sustained role that parents play in helping to shape and direct the character, sensibilities, and interests of their children" (p.197). This is something that children cannot do to their parents in return. Ivanhoe then suggests that the "only adequate way for children to respond ... is by living out of an attitude of loving care, appreciation, and reverence for their children" (p.197). Thus, one can "repay" the debt to one's parents by showing the same kindness, taking the same care and playing the same parental role towards one's own children.

All the three features of filial piety identified by Ivanhoe can be shown to have important environmental implications. However, while we can draw out the environmental implications of the Confucian commitment to filial piety, I will argue that it is possible to interpret the Confucian idea of filial piety in such a way as to show specifically that Confucianism entails a commitment to sustainable environment. Departing from the standard commentary on filial piety, I have elsewhere suggested that filial piety can be interpreted as ultimately respect, or reverence, for the tradition in which a person is embedded.[4] As we saw above, the way to be filial is to follow the rites, or rituals (*li*) (with the right attitude). Thus, while reverence, or respect, lies at the heart of filial piety, it is *li* that forms its structure. *Li*, in turn, is built up from past practices and thus *represents tradition*. To follow *li*, then, is to follow tradition. According to one commentator, "Confucian philosophy ... seems to find in tradition a reservoir of insight and truth."[5] Indeed, Confucius thinks of himself not so much as an innovator than as a transmitter of the wisdom embedded in a tradition going back to the sages of the past: "I am not one who was born with knowledge; I love ancient [teaching] and earnestly seek it" (*Analects* 7.20). Confucius declares that "in how to be a practicing gentleman, I can ... claim no insight" (*Analects* 7.33), and learns instead from traditional wisdom: "I believe in and love the ancients" (*Analects* 7.1). Elsewhere, the "Master said, 'the Zhou is resplendent in culture, having before it the example of the two previous dynasties'"

(*Analects* 3.14). Along the same line, Tu Wei-ming has claimed that human beings are, among other things, "historical beings, sharing collective memories, cultural memories, cultural traditions, ritual praxis, and 'habits of the heart'."[6]

If I am right then filial piety entails respect for tradition. Filial piety does not mean that one must respect one's parent as an independent figure of authority. Since a father has his own father and the latter his and so on, filial piety does not require the son to respect his father pure and simple, but rather to respect the father who respects his, and the latter in turn is someone who respects his, and so on. Filial piety is intergenerational. It makes no sense to speak of it as something that stops at one's own father. The father is really a father figure representing a tradition. Just as the wisdom of the Zhou culture was formed "having before it the example of the two previous dynasties," the views of the father, to which the son is meant to defer, are not entirely his own, but views formed "having before [him] the example" of previous fathers. Just as Confucius learns from and respect the ancients, a father must learn from and respect all the preceding fathers. This is why, contrary to the critics' claim that filial piety requires absolute obedience to the father from the son, it is possible, perhaps even necessary, for the son to correct the father when the latter errs, or holds view contrary to traditional wisdom. Thus, in Chapter 15 of the *Xiaojing*, we find that the Master denies that "filial piety consists in simply obeying the father," and insists instead that the filial son must "reason" with his father and "must never fail to warn his father against [moral wrong]."[7] In *Mencius* 6B3, it is said that while the parents' small fault may be tolerated, their big faults should be pointed out to them, out of our affection for them: "Where the parents' fault was great, not to have murmured on account of it would have increased the want of natural affection... [which] would have been unfilial." This is reinforced by Xunzi, who in Book 29 of the *Xunzi* on "The Way of the Son" (*Zidao*), asserts that filial piety requires getting a clear insight into the principle of when to follow and when not to follow the father's order, and applying this insight with courtesy and respect.[8] On my reading, one should follow the father's order when it is in line with traditional wisdom, and reason with him when his order deviates from it.

It may be said that reading Confucian filial piety as respect for tradition saddles Confucianism with conservatism. However, by "tradition" here I have in mind something close to Gadamer's understanding of tradition, according to which tradition is not a dead weight exerting its influence over an agent but rather a living consciousness with which an agent critically engages and to which he or she can contribute. As such, tradition is not simply backward-looking but also forward-looking. A living tradition is one that has the momentum to extend itself into the future. This is why Gadamer also speaks of tradition as a "horizon." Thus, respect for tradition entails playing a positive role to contribute to it and to carry it into the future. If filial piety means respect for tradition then filial piety entails the requirement to correct its defects as well as to preserve, nurture and enhance it and to help it extend into the future. Filial piety is not just backward-looking; it is also forward-looking. The forward-looking element of filial piety is perfectly consistent with the Confucian requirement to have children and to continue the family line. Thus, "Mencius said, 'There are three unfilial things [that a child can do] and to leave behind no posterity is the greatest'" (*Mencius* 4A26). However, it is not enough to have children and to continue the family line. As suggested above, to be filial is not to repay one's parents in kind for everything that they have done, which is not possible as Ivanhoe has pointed out, but rather it is to care for one's children, to make certain sacrifices for them, so that they will benefit in the same way as one has benefitted from one's parents' care and sacrifices.

I have shown that there is in Confucianism a strong commitment to filial piety as a key virtue, if not as the "root" of all virtues. I have shown also that this commitment is a commitment to a particular kind of future in which one's children and their children continue to exist. From this commitment, we can draw a number of environmental implications. The most obvious one is to extend the reverence, or respect, that one owes to one's parents as required by filial piety to nature itself insofar nature is regarded by

Confucians as our "parent."[9] However, in what follows, what I want to draw out of the Confucian commitment to filial piety is the commitment to ensure environmental sustainability. Indeed, in the next section, I will adopt Onora O'Neill's derivation of fundamental obligations to show that we can in fact derive from the Confucian commitment to filial piety a fundamental obligation to ensure environmental sustainability.

II

Confucian ethics is not an environmental ethics. To make it into one, we have to reconstruct it into an ethics that has the environment as its central concern and endorses specific obligations to the environment. It is not the aim of this paper to do so. However, I will argue that we can extract from the Confucian commitment to filial piety a fundamental obligation concerning the environment. In the case of non-anthropocentric environmental theories such as biocentrism and ecocentrism, which claim that nature has intrinsic value, the fundamental obligation rests on the intrinsic value of nature. Given the strong humanistic elements of Confucianism, the question is how to ground a fundamental Confucian obligation to the environment, if there is one. As it turns out, Onora O'Neill has attempted to ground a fundamental obligation to the environment in an anthropocentric foundation.[10] I will argue that we can follow O'Neill and construct a justification for a fundamental Confucian obligation to the environment. I turn now to O'Neill.

Given the rejection of anthropocentrism by supporters of environmental theories such as biocentrism and ecocentrism, the question is whether it follows that an environmental ethics cannot be strong unless it rejects anthropocentrism outright. O'Neill argues that it does not. Indeed, she claims that it will be preferable for an environmental theory not to have to suppose value realism, the view that nature has an intrinsic value that exists "out there" independently of valuers, and that an anthropocentric theory has this advantage. In particular, O'Neill endorses an obligations-based anthropocentric ethics. An obligations-based ethics is indeed strongly anthropocentric because only humans have obligations.

Nevertheless, it can serve as a foundation for a strong environmentalism if among the obligations that humans have are obligations to the environment, which require us to sacrifice some of our interests for the sake of the environment. According to O'Neill, such obligations have to be moral in nature and thus they have to be fundamental obligations rather than institutional. Being fundamental, they must be based on principles that are not just universalizable but also can "be accepted and adopted (not necessarily discharged) by all agents" (p.135). A universalizable principle (e.g. taking the bus to work) will not be accepted and adopted by all agents if any of its alternatives (e.g. taking the train to work) is also universalizable (O'Neill's own example: fasting by day and fasting by night). However, when a principle is universalizable and none of the alternatives is then the universalizable must be adopted as obligatory (and any alternative must be rejected). For instance, a Kantian might say that since telling the truth is universalizable and its (only) alternative, lying, is not, it is obligatory to tell the truth, or obligatory not to lie.

O'Neill goes on to identify a principle that serves her purpose, namely the principle of rejecting the commitment to injure because it is universalizable and its only alternative, the commitment to injure, is not. For O'Neill, rejecting the commitment to injure is a fundamental human obligation. To discharge this obligation, a human society must develop institutions designed to minimize injury to persons. Since we can be injured by living in a degraded environment, we have an obligation not to damage the environment:

> ...if the rejection of systematic or gratuitous injury to other agents is a fundamental obligation, then it will also be obligatory not to damage or degrade the underlying powers of renewal and regeneration of the natural world. The basic thought here is that it is wrong to destroy or damage the underlying reproductive and regenerative powers of the natural world because such damage may inflict systematic or

gratuitous injury on some or many agents (p.137).

Since the "reproductive and generative powers of the natural world" are not confined to sentient creatures, our obligations extend beyond them. From here, we can argue for an extensive and far-reaching protection of the natural world. For those aspects of nature that still remain out of reach, we can establish "…indirect arguments deriving from human obligations for … [a] … wider protection, *or* … choose to establish positive obligations" for such protection (p.139). All this, O'Neill claims, should be sufficient to endorse the perspective of strong environmentalism and should thus satisfy strong environmentalists.

It is likely that strong environmentalists would not be satisfied with O'Neill's fundamental obligation "not to damage or degrade the underlying powers of renewal and regeneration of the natural world." However, O'Neill's account remains useful insofar as it is not my intention here to argue for the claim that Confucianism can furnish the ground for a strong environmental ethics (although I have argued elsewhere that it can).[11] Rather, my aim is simply to argue that Confucianism is committed, at a minimum, to ensure a sustainable environment. With this modest aim in mind, we can first agree with O'Neill that "it is wrong to destroy or damage the underlying reproductive and regenerative powers of the natural world because such damage may inflict systematic or gratuitous injury on some or many agents." Next, we can supply a Confucian account of why it is wrong. To accomplish this, we need to supply a Confucian alternative to O'Neill's fundamental moral principle, to her principle of rejecting the commitment to injure.

Given the Confucian commitment to filial piety discussed in Section I, the following principle can be proposed as a Confucian principle:

C = We ought to ensure the flourishing of the institution of the family.

C, arguably, is what the Confucian commitment to filial piety amounts to. As pointed out above, it is not just a matter of revering, or respecting, one's own parents. Filial piety has to be understood as respect for tradition, in particular, for the family tradition as a living institution. To be filial is to be grateful for the parental care and sacrifice made by one's parents and by all the preceding parents, as well as to show the same care and make the same sacrifice for one's children and their children. In filial piety, we look back into the past to assess our debt to it, and forward into the future to determine what we owe to future generations. As noted above, human beings are historical in the Confucian visions of human beings, according to Tu Wei-ming, and it is mainly through exhibiting the virtue of filial piety that we are historical. If "leaving behind no posterity" is, as Mencius says, the most unfilial thing to do then it must also be the most unfilial thing to leave behind a kind of posterity that is so impoverished as to be unable to leave a further posterity behind it. Since our posterity needs certain minimum environmental standards to survive, it is the most unfilial thing not to ensure such standards. Indeed, we can simply borrow O'Neill's words to characterize *C* as a principle that requires us not "to destroy or damage the underlying reproductive and regenerative powers of the natural world because such damage may inflict systematic or gratuitous injury on" future generations to the point where they are unable to flourish, or worse still unable to regenerate. We can conclude then that *C* is a principle derivable from the Confucian commitment to filial piety. The question is whether it is a *moral*, that is, *fundamental* principle. It might be sufficient to argue that since filial piety is a fundamental virtue and the commitment to it is a fundamental moral requirement, anything derivable from it, such as *C*, must be fundamental. However, we can employ O'Neill's strategy to show that it is indeed fundamental: we can show both that *C* is universalizable and that *non-C* = It is not the case that we ought to ensure the flourishing of the institution of the family, is not.

That *C* is universalizable and *non-C* is not is fairly straightforward. As respect for tradition, filial piety is not a commitment just to one's own family. Even if the primary focus is on one's own family, it is still the case that no single family, like no single individual, can flourish without the community as a whole flourishing. Much has been said of the Confucian conception of the self as a being standing in a network of social relationships rather than an individual independent of others in the society. What is true about the self is also true about the family. Thus, just as there is no self without others, there is no family without other families. For any family to flourish, the whole family institution has to flourish. It follows that to be filial is to commit to C. Since filial piety is a key virtue, if not the most fundamental virtue, all moral agents have to cultivate it. It follows that C is universalizable and *non-C* is not. This is the case at least as far as Confucianism is concerned. Following O'Neill, we can say that C is a fundamental moral principle in Confucianism. Since C entails the obligation to sustain an environment suitable for the flourishing of the family institution, environmental sustainability is a Confucian fundamental moral obligation.

Could C be a fundamental moral principle for those who do not subscribe to Confucianism? It is arguable that the emphasis on the family is not confined to the east. The family has been a key institution in the west and remains alive and well today. In most if not all environmental ethics, the continuation of the human species is taken for granted as desirable. Insofar as the family unit is crucial to the continuation of the human species, any commitment to the latter requires a commitment to the former. If so then C may be said to be universally universalizable. The question now is whether *non-C* is non-universalizable outside of Confucianism. Given the different life styles in the west, some of which are not based on the family unit, it seems that *non-C* is also universalizable, as much as taking the bus (rather than the train), or fasting in the morning (rather than the evening). However, while not taking the bus (or train) does not cause the collapse of the transport system, and while not fasting in the morning (or at night) does not adversely affect the practice of fasting, as long as it is admitted that the family unit is crucial to the continuation of the human species and as long as it is believed that the continuation of the human species is desirable, *not-C* is not universalizable even for those who choose a life style not based on the family unit. Thus, it seems that the Confucian argument for the claim that there is a moral obligation to sustain a livable environment indefinitely into the future is universally sound as long as the institution of the family (or the family unit in biological terms) is necessary for the survival of the human species and as long as the latter is desirable in itself.

III

As mentioned above, Confucian ethics is not an environmental ethics and the obligation to achieve a sustainable environment is not directly stipulated in Confucianism. However, I have argued that we can derive such obligation from Confucianism. Nevertheless, such derivation must have some textual basis. The question is whether there is textual evidence to show that prominent Confucians advocate sustainable practices as well as ideas and practices that contribute to or result in sustainability. The answer to this question depends, in part, on what is meant by environmental sustainability, or sustainable development. Unfortunately, there is no general agreement on either the definition or the measurement of sustainability. However, for our purpose here, we can adopt the definition found in the United Nation's Brundtland Commission Report issued in 1987, namely: "Sustainable development is development that meets the needs of the present generation without compromising the ability of future generations to meet their needs." The definition seems to agree with the Latin etymology of "sustainability," which is *sustinere* (a combination of "hold" (*tenere*) and "up" (*sus*)). Thus, if we are able to meet our needs, we will be able to "hold" ourselves "up," that is, to maintain ourselves and to extend the human community into the further future. The process of meeting our needs is what is meant in this context as "development." Many commentators claim that there are different spheres of development and that sustainability refers to sustainability in all these different spheres. The spheres are commonly taken to be the social, the economic and the environmental spheres. Social sustainability refers roughly to the sustainability of the key social institutions that enable humans to "hold" themselves "up." For Confucians (and arguably for all others), the most important of such institutions is the family. Given the discussion above, we can say that there is a direct Confucian commitment to social sustainability, which can be summed up in Mencius' remark, cited above, that the most unfilial thing to do is to leave behind no posterity.

Turning to the other spheres of sustainability, it can be argued that there is in Confucianism an indirect, or derived, commitment to economic and environmental sustainability. Given the strong Confucian commitment to social sustainability, it would be surprising if there is no commitment to economic and environmental practices that contribute to it. The economy and the environment have to be sufficiently developed, and the development sustainable, so as to enable human societies, present as well as future, to meet their needs. In this context, it is useful to remind ourselves that Confucians place a great of emphasis on the idea of harmony ("Of the things brought about by the rites, harmony is the most valuable" – *Analects* 1.12). Without being able to meet our needs, there can be no harmony. Without harmony, there can be no posterity. Thus, filial piety entails a commitment to ensure that future generations have enough to meet their needs so that they can maintain social harmony. Whether future needs are adequately met depends on a number of factors. Adapting a well-known formula in environmental science according to which the impact on the environment (I) is a function of population growth (P), the level of affluence (A) and the technology employed to meet our needs (T) -- $I = P \times A \times T$ --[12] we might say that the impact on the ability of future generations to meet their needs depends on population growth, consumption patterns and productive technology. To be committed to posterity is to be committed to maintaining a proper mixture of these three variables. Regarding P, there is no textual evidence to show that Confucianism is committed to any particular population policy other than the requirement to have children. However, there is ample textual evidence to show that Confucianism advocates consumption and production practices that contribute to economic and environmental sustainability. Thus, while following Confucian practices may not affect the variable P, it will favourably affect the other variables, namely A and T.

Everything being equal, the higher the level of consumption, or affluence, pursued by the present generation, the greater will be the demand on scarce resources and the adverse impact on the environment, both of which reduce the ability of posterity to meet their needs. It follows that if the present generation is frugal in its consumption practices, or curtails its appetite for affluence, future generations are more likely to be able to meet their needs and to sustain themselves. In this respect, Confucianism is well known for its advocacy of frugality. While it is not clear whether Confucius did in fact say "He who will not economize will have to agonize," a statement commonly attribute to him, he certainly advocates economizing and moderation in consumption. At *Analects* 9.3, Confucius says: "The linen cap is that prescribed by the rules of ceremony, but now a silk one is worn. It is economical, and I follow the common practice." It is not just individuals that are urged to be economical, Confucius also counsels governments to be economical. Thus, at *Analects* 1.5, he says that to "rule a country of a thousand chariots, there must be," among other things, "economy in expenditure." While the rites are important, Confucius advises against being extravagant in following them: "In festive ceremonies, it is better to be sparing than extravagant" (*Analects* 3.4). The rejection of extravagance is repeated at *Analects* 7.35: "Extravagance leads to insubordination, and parsimony to meanness. It is better to be mean than to be insubordinate." The rejection of extravagance is consistent with the teachings in the *Doctrine of the Mean*, a work that begins with a remark according to which the *chung* in *Chung Yung* means "without inclination to either side." Thus, over-consumption is to be avoided. Indeed, the bias is towards under-consumption. At *Analects* 6.9, Confucius praises Hui for his simple tastes: "With a single bamboo dish of rice, a single gourd dish of drink, and living in his mean narrow lane ..., he did not allow his joy to be affected by it. Admirable indeed was the virtue of Hui." In the same way, Confucius praises Yu: "I can find no flaw in the character of Yu. He used coarse food and drink, but displayed the utmost filial piety towards the spirits. His ordinary garments were poor, but he displayed the utmost elegance in his sacrificial cap and apron. He lived in a low mean house, but expended all his strength on the ditches and water-channels" (*Analects* 8.21). Confucius himself emulated Hui and Yu: "With coarse rice to eat, with water to drink, and my bended arm for a pillow; I have still joy in the midst of these things" (*Analects* 7.15).

Compared with Confucius, Mencius does not have as much to say about frugality. However, at 7B34, he expresses a strong preference for it: "Their hall is tens of feet high; the capitals are several feet broad. ... Their tables, laden with food, measure ten feet across, and their female attendants number in the hundreds. ... I would not indulge in such things." At various places, Mencius speaks against waste. Thus, at 6B11, he says: "Inundating waters are a waste and what a benevolent person detests." At 7B34, he advises against building and living in big mansions, which is a waste of timber.

Turning to the final variable in the sustainability equation above, namely T, or sustainable technology, again, there is ample textual evidence to show that Confucius and particularly Mencius advocate the production practices that would maintain the "underlying reproductive and regenerative powers of the natural world," to borrow O'Neill's words. Thus, at *Analects* 7.26, we are told that when Confucius fished, he used a line and not a net so as not to affect the fish stock, and when he went shooting, he spared roosting birds so as not to affect future supplies. As for Mencius, several passages from the *Mencius* can be cited to show that he was well aware of the importance of the "reproductive and regenerative powers" of nature. At 6A8, in the context of discussing human nature, Mencius laments the state of Ox Mountain, the trees on which were cut down by villagers and the regenerated young plants were eaten by grazing animals. Like the goodness of the human heart, the vegetation of the Ox Mountain has to be "given the right nourishment," without which "nothing ... will ... grow." At 1A3, he gave the following advice to King Hui: "If you don't interfere with the timing of the farmers, there will be more grain than can be eaten. If fine-mesh nets are kept out of the ponds and lakes, there will be more fish and turtles than you can eat. If loggers are regulated in their woodcutting, there will be more wood than can be used." This is followed by the advice to increase the productive base to ensure future consumption at 1A7: "If mulberry trees are planted around homesteads of an acre, then people fifty years old can be clothed in silk. If, in the raising of fowl, pigs, dogs and swine, their breeding times are not missed, then people seventy years old can eat meat. If you do not upset the farming schedule in a farm of twenty acres, then a large clan will never be hungry." To maintain the "reproductive and regenerative powers" of the land, Mencius, at 6B7, advises the government to reward officials who nourish old fields and punish those who neglect them.

I have argued that the Confucian notion of filial piety should not be understood narrowly as respecting one's own parents; rather, it should be understood widely as a moral commitment to a historical tradition that extends backward to one's ancestry and forward to posterity. In particular, I have argued, following O'Neill's procedure for identifying a fundamental moral principle, that Confucian filial piety entails the moral commitment to the sustainability of the institution of family, which in turn entails the moral commitment to environmental sustainability. I have shown that the derivation of the fundamental moral obligation to ensure environmental sustainability from Confucianism can be supported by the fact that there is ample textual evidence to show that the early Confucians advocate consumption and production practices aimed at ensuring sustainability. Indeed, one commentator goes so far as saying that the present generation of the Chinese people would be better able to meet their needs if "the Chinese rulers and people had heeded Mencius' advice."[13] If this is true then it is not unreasonable to assume that by heeding Mencius' (and Confucius') advice now, we will "meet the needs of the present generation without compromising the ability of future generations to meet their needs," which is what is commonly understood as sustainability. If the practices advocated by Confucius and Mencius turn out to be no longer efficacious, we need to discover practices that are, because there remains the fundamental moral obligation to ensure environmental sustainability.

NOTES

1. Translations of the *Analects* have been adapted from various sources.

2. Translations of the *Mencius* have been adapted from various sources.

3. Philip J. Ivanhoe, "Filial Piety as a Virtue," in Alan K.L. Chan and Sor-hoon Tan (eds.), *Filial Piety in Chinese Thought and History* (London and New York: RoutledgeCurzon, 2004), pp.189-202.

4. A.T. Nuyen, "Filial Piety as Respect for Tradition," in Alan K.L. Chan and Sor-hoon Tan (eds.), *Filial Piety in Chinese Thought and History* (London and New York: RoutledgeCurzon, 2004), pp.203-214.

5. Alan K.L. Chan, "Confucian Ethics and Critique of Ideology, "*Asian Philosophy*, Vol.20(2000), pp.245-261, at p.245.

6. Tu Wei-ming, "Beyond the Enlightenment Mentality," in Mary Evelyn Tucker and John Berthrong (eds.), *Confucianism and Ecology: The Interrelation of Heaven, Earth, and Humans* (Camb., Mass.: Harvard University Press, 1998), pp.3-21, at p.13.

7. Translation of the *Xiaojing* has been adapted from M.L. Makra, *The Hsiao Ching* (New York: St. John's University Press, 1961).

8. Translation of the *Xunzi* has been adapted from J. Knoblock, *Xunzi: A Translation and Study of the Complete Works* (Stanford: Stanford University Press, 1994).

9. See, for instance, Tu Wei-ming, The Continuity of Being: Chinese Visions of Nature," in Mary Evelyn Tucker and John Berthrong (eds.), *Confucianism and Ecology: The Interrelation of Heaven, Earth, and Humans* (Camb., Mass.: Harvard University Press, 1998), pp.105-121.

10. Onora O'Neill, "Environmental Values, Anthropocentrism, and Speciesism," *Environmental Values*, Vol.6(1997), pp.127-142.

11. A.T. Nuyen, "Confucian Role-based Ethics and Strong Environmental Ethics," *Environmental Values*, forthcoming.

12. P.R. Ehrlich and J.P. Holden, "Human Population and the Global Environment," *American Scientist*, Vol.62, No.3(1974), pp.282–292.

13. J. Donald Hughes, *An Environmental History of the World* (Abingdon, Oxford: Routledge, 2009), p.71.

CHAPTER NINE

Toward An Ethical Climate Regime*

Po-Keung Ip

Introduction

Anthropogenic climate change has produced dangerous consequences that posed serious threat to human society (IPCC, 2007). World leaders have been feverishly working together to find solution and implement collective action to combat these threats and dangers. The harm-causing consequences of a warming world are more evident in some areas than others. Some poor and small island nations have already suffered from the damages caused by adverse climatic change, while many developing countries are becoming more vulnerable to the immanent threats – severe floods and droughts, rising sea levels, intense storms, increased heat waves, diseases and deaths. Intricately meshed with these harms created is ethics. Who is responsible for creating these harms and dangers? What ought to be done? These are some of the pressing ethical issues humanity need to address in response to climate change. Other ethical issues like fairness of allocation of emission rights, justice in mitigation and adaptation, compensation justice in rectifying past wrongs in creating climate harms etc. are integral to climate change problem that are critical and unavoidable.

Rich nations undoubtedly are mostly responsible for the buildup of greenhouse gases (GHGs) in the atmosphere to present levels, though the total emission and per capita emission levels vary greatly among nations. Those who are harmed and get benefit (from GHGs emission) are located in both poor and rich nations. However, those who are most vulnerable to climatic harms are often the least responsible for the emissions. As well as unfairly shouldering the climatic burden, they are often the least able to adapt, like building dikes, irrigation to compensate for droughts, and moving away from flood or storm prone areas. Lacking both financial and technological means, they are the most ill-equipped to meet the challenges posed by climatic change.

In response to the climatic threat, international negotiations began in the late 1980s, with some nations and organizations calling for enforceable targets. These agreements include the Rio Declaration on Environment and Development, United Nations Framework Convention on Climate Change (UNFCCC) (Framework Convention hereafter) with the Kyoto Protocol set out norms, measures and targets to combat the issue. Like other major global environmental issues, the climate change issue can be seen as a major issue in the governance of the global commons. Humanity designs and develops governance regimes to handle these issues. Climate change is inseparable from ethics. A successful governance regime should be ethical as it should be effective. Ethics give moral legitimacy to the regime. Effective regimes have binding norms to coordinate cooperation and guide collective actions. Taking the Framework Convention and the Kyoto Protocol as a governance regime for climate change, I report the obstacles challenging global commons governance, note the strengths and the weaknesses of the Kyoto protocol, identify the ethics inherent in climate change and formulate some major principles of an ethical climate regime through which an ethics-based Climate Treaty could be framed.

Global Commons Governance and Its Obstacles

To effectively respond to climatic threats and problems, effective governance regimes in addition to analytical framework for understanding the complex issues are needed (Dietz, et al. 2003; Chasek et al, 2006; Ostrom, 2009). A regime is a system of norms, measures and institutions to coordinate and regulate collective behaviors of actors to find solutions to a common problem. A regime is, (Chasek et al. 2006: 17),

"a system of principles, norms, rules, operating procedures, and institutions that actors create or accept to regulate and coordinate action in a particular issue area of international relations."

Global commons issues are complex and transcend national borders. They involve multi-stakeholders with diverse interests, and embedded in a long time horizon with varied impacts and costs across nations and groups. To fashion and execute a collective response to these issues requires the regime to coordinate, regulate and direct activities and policies towards a defined objective. The regime's values, norms, policies, measures, mechanisms, and institutions work together to serve these functions. Norms guide, coordinate and regulate participants' behaviors, and they also define the related responsibilities for them. As noted earlier, the UNFCCC with the Kyoto Protocol is an example of a climate change regime. Likewise, other major efforts in addressing global environmental problems have also been executed through regimes which include international treaties, norms, principles, measures, targets and financing and implementation mechanisms. The standard view of a regime does not include the element of agency - agents who play crucial parts in the cooperative activities. I supplement the standard view by including moral leadership as a crucial constituent of the regime. Past experiences in regime design and implementation can provide insights and valuable lessons on how to develop effective regimes for climate change.

Dietz et al. (2003) has recently proposed some conditions for an effective commons governance. These include, first, the resources of the commons and the human users of the resources can be monitored, measured and verified at relatively low cost; second, there are moderate rates of change in resources, resource-user populations, technology, and economic and social conditions; third, there is sufficient social capital among the communities involved for rule compliance; effective monitoring and the enforcement of rules are supported by users communities, among others. Few systems in the real world can meet all these conditions. This fact helps to highlight the challenges of commons governance.

There are structural and procedural obstacles facing global commons governance. (Chasek *et al.* 2006). The structural obstacles include the structure of international systems, international law, and the global economic system. On the procedure side, the process of negotiation and decision-making on the issues of the global commons creates hurdles. Also, some features of global commons introduce difficulties. These three kinds of obstacles are discussed next.

One conspicuous structural factor apparently responsible for many past failures in providing effective solutions to global issues is the failure of nation states to form a strong super-state to govern global affairs. Under the present political arrangement, effective and durable cooperation is difficult because of the unequal power and influence of different member states. The wills of strong states often dominate the game of cooperation and terms of negotiation, exasperating fear of an unfair game of cooperation on the part of the weak states. Other difficulties include the failures to deal with the problems of non-compliance, free riding, mutual suspicion, lack of trust, misperceived motives or actions of other members. The second structural obstacle is produced by the current political system of decision making which is dictated and driven by national interests with short time-frames. This system is acutely ill-suited to handle global commons problems that respect no national boundaries and involve long time-frames. The solutions to global commons issues like ozone depletion, fishery stocks depletion, cross-border pollution require a mandate and execution that transcend the function of current political systems that gives sovereign states exclusive control over their territories. In the global commons, however, what one state acts within one's border can effectively harm others or the commons.

On the procedure side, the lengthy and tortuous processes of negotiations are intensively contentious and time-consuming, inadvertently creating a time lag problem. One famous example is the Basel Convention on transportation of hazardous waste across borders. Though the problems were identified in the 1970s, the Basel Convention was effective not until 1992, and its Ban Amendment is still awaiting additional ratification. Other areas in the global commons, like protecting endangered species also are plagued by the same problem. While negotiating, species become extinct, and biodiversity declines. In the case of the Kyoto

Protocol, while still seeking ratification, the world is getting warmer. The other procedure hurdle is commonly known as the "lowest common denominator problem". The fate of treaties addressing global commons problems can be effectively dictated by the least cooperative states, the so-called 'veto-states'. These states can possess and exercise 'veto power' over terms of the treaty according to their wishes. The trade-offs that made to secure their participation may compromise the regime's integrity and effectiveness. One example of veto-states playing mischief is the global ozone treaty. The European Community from 1977 to 1989 played the role of a veto state by successfully defining the weakest terms of the treaty, effectively weakening the provisions in the 1985 Vienna Convention and the 1987 Montreal Protocol.

Added to these two obstacles are the hurdles created by the nature of the global commons (Chasek et al. 2006, 204-208). Issues in the global commons are intimately connected to vital economic and political interests of nations and groups, which usually trump global commons concern. For example, to control the emission of GHGs requires a basic change or modification of energy policies of many fossil-fuel based economies that entails heavy adjustment costs. National differences in resource endowment and energy system (e.g., coal-oil based vs. nuclear-based) create unequal adjustment costs for different nations that spawn the formation of veto coalitions to protect their interests. Because many of the grave consequences of global commons issues take a long time to emerge, nations will shy away from supporting strong regimes for these remote consequences, thanks to the short-termism of the present-day politics. These hurdles will be even made more insurmountable if the issues in hand are intractably complex and scientifically uncertain, amplified by a lack of concern of the electorate as a result of these complexity and uncertainty. Past experiences in the control of biodiversity loss and ozone depletion have confirmed this. Furthermore, a large number of participants render effective cooperation difficult because of the diverse interests, motivations, and perceptions entangled in the process. Lacking a strong monitoring authority increases the chances for participants to cheat and free ride. Furthermore, a kind of "use it or lose it" mentality that the situation created will exasperate the problem. The depletion of ocean fisheries is a classic example.

The Kyoto Protocol – Its Strengths and Weaknesses

The Framework Convention was signed in 1992 and became effective in 1994. It provides a master norm that defines in general terms how collectively to respond to the climate problem. To translate this norm into action spawned the Kyoto Protocol in 1997 with specified quantitative targets to achieve the Framework's objective, and set the limit for industrial nations for reducing greenhouse gas emissions, about on average 5 percent below 1990 levels over the 2008-2012 period. Industrial nations began resolving implementation details and began to ratify the Protocol. The Bush Administration of the US rejected the protocol by not ratifying it. However, with the Russian ratification, the Protocol came into effect in 2005. As of today, 184 Parties of the Convention have ratified its Protocol. The implementation details of the Protocol, called the "Marrakesh Accords.", were adopted in Marrakesh in 2001.

The Protocol allows the parties to the agreement to achieve their targets through national means, but also offers market-based mechanisms – a. emission trading ("carbon market"); b. clean development mechanism; and c. joint implementation. The mechanisms are expected to facilitate investment in green technologies and development of cost-effective means to achieve their targets. Furthermore, the monitoring system is set up to monitor actual emissions of the parties and to keep records of the emission trading. The parties are required to submit annual reports and emission inventories to the secretariat. There is also a compliance system to ensure that parties are meeting their commitments. In addition to the mitigation means, the Protocol has means to deal with adaption problems by facilitating the development and utilization of technologies to enhance the capacities of the parties to meet the challenges posed by climate change. Despite many criticisms, the Kyoto Protocol is widely regarded as an important first step towards solving the climate change problem. In 2012 when the protocol expires, a new post-

Kyoto climate treaty will be negotiated in Copenhagen as a continued effort to combat the challenges of the climate change.

Though far from problem-free, the significance of the Kyoto Protocol should not be under-estimated (Aldy and Stavins 2008: 8-9). It is a major global undertaking, through intense, costly and tough negotiations, to consolidate a collective response to the threat of climate change. The Kyoto regime developed over time, including the various rules, ways and means, and institutions to combat GHG emission, is flexible and innovative, as well as pragmatic and actionable. It also provides a solid foundation for further strengthening and development, as well as modifications in future agreements. With its principle of common and differentiated responsibilities, the Protocol recognizes the issue of allocation fairness by allocating larger responsibilities to rich and industrial nations because of their past deeds in contributing to the accumulation of GHG in the atmosphere. In addition, it also calls for those rich nations to shoulder a heavier burden to pay for the mitigation efforts. As a global commons governance regime, it seems to have a design architecture that broadly responds to some obstacles mentioned earlier, as far as it is practicable. Other obstacles like compliance, are inherent difficulties that defy quick fixes, but demand long-term solutions.

Notwithstanding its design niceties, one major weakness of the Protocol is its failure to include the participation of China and India, the top GHGs emitters into the process (they do not have emission targets), while allowing Russia to adopt very lax targets. More importantly, the United States, being the largest emitter, refused to ratify the protocol. The combined emission of these four top emitters contributed half of the global emission in 2004. Their share will expect to grow in view of the rapid growth in China and India. As a result, the potential gains from other parties complying with the targets and from trades are effectively lost. Also, the cost differentials across nations have motivated carbon-intensive corporations in nations with emission commitments and high costs to relocate some operations to nations with lower costs and without emission commitments. This gap nullifies benefits from the mitigation effort of the rich parties to the agreement and weakens participation.

Ethics of Climate Change

The Framework Convention and the Protocol both clearly acknowledge one basic ethical principle that underlies the global response to climate change. The principle requires that the cooperative response of the members of the global communities should be in line with 'their common and differentiated responsibilities and respective capabilities and their social and economic conditions'. This principle has strong ethical undertone as the stated responsibilities embrace both the legal and ethical senses. This principle however only partially represents what is needed to sustain an ethics-based climate regime. In addition to this principle, the post-Kyoto climate regime should embrace a more robust ethic that inform, guide and regulate strategies, policies and practices. Perhaps it was as the result of the tough and contentious negotiations dictated by power politics that left little room for a fuller expression of ethical concerns. Furthermore, the predominance of the legalese wordings of the provisions and articles of the agreement might have effectively suppressed more explicit articulation of ethical concerns, and inadvertently pushed ethics to the margin. To do full justice to ethics, a preliminary unpacking of the ethics involved in climate change is in order.

In the most basic sense, ethics demand people avoid causing harm to others, or seek to promote happiness and well-being of others. One major function of ethics is to respect the rights of others by not violating or infringing them. It is a truism in morality that people ought not to be harmed, and people's rights ought not to be violated. People have basic rights to life, health and security. They have basic moral right not be harmed in these major aspects. The adverse consequences of anthropogenic climate change threaten and infringe these rights, and causing serious harms to people. Those responsible for the harm caused or the violation should be held accountable and made to compensate for the harm caused or rights violated. There is also a major justice issue regarding sharing benefits, risks and burdens relating to climate change. A critical issue of allocating emission rights to members of the global community, indeed, stands at the core of the ethics of climate change. Who should have how much of these rights and why? It is clear that the major issues of human-induced climate change are intimately connected with ethics (Broom, 2008; Shue, 2001; Gardiner, 2004; Raymond, 2008; Brown et al 2008; Singer, 2002; Vanderheiden, 2008).

Collective norms have been emerged and affirmed over the long process of negotiating common responses to the global commons issues in the last couple of decades. The Rio Declaration and UNFCCC include norms that demand nations to have the responsibility to reduce GHG emissions of activities, to have the responsibility to reduce their emission based upon equity considerations, the developed nations have the responsibility to assume leadership role in reducing the threat of climate change. As said, the issue of justice is an integral part of climate change. Retributive justice should guide the attribution of responsibility proportionate to those responsible for harm-causing GHGs emission activities. Distributive justice should regulate the allocation of emission rights of nations. In addition to these norms that define what ought to be done, there are also norms that discourage what ought not to be done. For example, nations should not use scientific uncertainty as an excuse for not taking actions; nations should not use cost to national economy as an excuse to shirk responsibility to reduce emission. Though articulated as treaty provisions, these norms are ethical in nature as well.

An Ethical Climate Regime – Its Principles

I outline some major ethical principles that help define an ethical climate governance regime. As stated earlier, a governance regime not only has to be effective, it has to be ethical. An ethical regime possesses the following basic features: 1. It helps define ethical norms, strategies, policies, measures, and mechanism. 2. It helps generate and support ethical norms, strategies, policies, measures or mechanisms; 3. It helps protect the integrity and well-being of the global commons; 4. It is based on legitimate basic values that are compatible with universal values, supportive of the well-being of humanity, and is not based on narrow or special interests; 5. Its norms, policies, measures and mechanisms are compatible with the universal moral demands of non-harming, justice, and basic human rights, and the caring for the well-being of non-human species and the ecosystem. With these features, the strategies, norms, policies, measures, and mechanisms informed, formulated or adopted through the ethical climate regime are ethical, or have positive ethical consequences. The seven proposed principles include the principles of sustainability, common and differentiated responsibilities with respective capacities, allocation justice, retributive justice, obligation to protect, inclusiveness with procedure justice, and moral leadership. Though not forming an exhaustive set, these principles serve as necessary conditions of the ethical regime.

Sustainability

The climate response should follow the principle of sustainability which requires that development and human activities should meet the needs of the present generation without depriving the benefits and well-beings of future generations and other species sharing the same earth with humanity. The concept of sustainability can be articulated in terms of the flow of stocks in natural and human capitals (Turner and Pearce, 1993). Specifically, sustainability refers to a situation where the stocks of natural and human capitals can be maintained in a non-declining state. The ethics of sustainability require the continuous maintenance and nurturing of stocks of natural and human capitals for humanity's continued survival and well-being (Ip, 2004). Natural capital and human capital respectively refer to the natural resources and human and social resources that sustain human activities and flourishing. The ethics of sustainability embrace not only cross-generation justice, but also cross-species justice. Thus it is a deeply ethical principle applicable across generational as well as species divides. Climate change takes place over a long time frame which renders policies framed in short-termism futile and ineffective. Long-term, intergenerational, and cross-species sustainability is what a durable ethical climate regime requires.

Common and Differentiated Responsibilities with Respective Capacities

The present climate crisis is the result of years of industrialization of the developed economies of world. These fossil-fuel-based economies, though have brought unprecedented wealth to the rich nations and lifted millions out of the poverty, are unsustainable and are the main culprit of the present climate menace. Developed economies have reaped enormous economic benefits from the process while creating heavy climatic burdens for themselves and others by emitting earth warming GHGs over the globe. Great benefits gained by the rich nations have not been equitably balanced with their proportionate burden-sharing activities. Innocent parties, especially developing nations which have played negligible roles in atmospheric GHG accumulation however, are facing increasing and imminent threats from climatic change. This creates an acute problem of distributive justice.

The global nature of climate change demands a global response. It means that every nation, including everyone living on this planet, has to share the responsibility to respond. But due to differences in past deeds, those who have contributed more to this problem should bear a larger share of responsibility. This means that rich developed nations should bear more responsibilities than those of the poor developing nations. By virtue of past harm-causing wrongs committed by some parties, these parties should bear the responsibilities for righting these wrongs. Fairness requires that responsibilities in climate change responses among nations should be differentiated, and not equal, in accordance to their varied historical acts. Notwithstanding the complexity and difficulties in identifying and measuring how much harm has been caused, the principle of paying for one's debt is a reasonable moral demands in attributing responsibilities.

The attribution of responsibilities not only has to take into account of the past deeds of the parties, but also their respective capacities to response. It is not right to assign responsibilities to parties who lack the capacities to stage and sustain an effective response. The Humean dictum of ought implies can works here. Only those who are able to effectively respond to the climatic threat should be allocated the related responsibilities to respond. Lacking the respective capacity can exempt nations or groups from shouldering responsibilities. To enable every nation or group to share the common responsibilities, rich and able nations should help build related capacities of those who lack them. In treaty terms, it means that developed nations should help developing and poor nations through financial and technological means to build up their capacities in both mitigation and adaptation against climatic threats.

Allocation Justice

One key task of an ethical climate regime is to solve the "who get what and why" problem of fairly allocating emission rights. The principle of allocation justice is purported to decide fairly who gets how much by way of emitting GHGs, taking the atmosphere as a carbon sink, according to the related fairness principle which allows for different formulations. Fairness can be formulated in terms of per capita, per nation allocation, or in terms of allocation over industries or sectors. Each formulation of fairness may have different implications for different nations regarding the amounts of emission rights, among others. Per capita allocation may benefit developing nations more as they are more populous than the developed nations. Allocation on per nation basis may disadvantage populous nations and give rich and less populous nations more emission rights. Allocation fairness should also take into account of the past deeds of parties that have contributed to the problem. As historical acts of nations varied, some kind of proportionate allocation should be designed, provided the impacts of past activities can be reasonably identified and measured. These issues have to flesh out in details to give content to the principle of allocation justice.

Retributive Justice

The polluting acts of the industrial nations over the past decades could be seen as acts of heavy borrowing from the global commons. They are heavily indebted to other non-polluting members. Justice demands that debts have to pay, especially to those not or least responsible for the mess of climatic indebtedness. Compensatory justice requires that debt-bearing nations discharge the responsibility of paying back past debts owed to others. In practice, it means rich nations should help poor nations by financial means or technological support in building and developing their capacity to protect themselves from climatic harm.

Obligation to Protect

The impact of adverse effects of climate change varies. Some nations will be hit harder than others. Some nations will suffer more than others. For those nations least able to cope with the threats and potential harms, like many small Island states and poor nations, the grave consequences of climate change may pose fatal risks to the very survival of their inhabitants. Hundreds of millions people's lives and dwellings are under real threat of annihilation and destruction. In particular, some small tribes in the severely affected areas may even face the threat of extinction. In this aspect, climate change is deeply an issue of human rights as well. At stake here is the weakest and most vulnerable people's right to life and their right to survive. Ethics require that people's basic rights should be protected. People who are able should have an obligation to help those in dire need or danger. In particular, those more able members should have a special obligation to protect the weak and the vulnerable. An ethical climate regime should take this obligation to protect seriously in framing stratgeis, norms and policies.

As well as affirming differentiated responsibilities, the regime should define differentiated responsibilities to protect the weak and most vulnerable. The climatic change impact on nations varies, and nations also vary in their capacity to respond to the threats. Some nations (e.g., Maldives) are more vulnerable than others to the threats and harms, and some nations (e.g., the Netherlands) are more capable than others to protect themselves from the threats and harms. Ethics demand that the able should help the less able, the strong should aid the weak. Morality requires that the weakest and most vulnerable deserve more of our concern and intervention. Wealthy and able nations should have the obligation to protect those nations most vulnerable to climatic harms. Coupled with the fact that poor nations are the least responsible for the harm caused to the atmosphere, it makes compelling case that rich economies should have a special obligation to protect.

Inclusiveness and Procedure Justice

As the varied impacts of climate change affect everyone on the planet, no one can escape from them. Global impact calls for global response. The climate regime should include all the stakeholders into the process of negotiation and formulation of policies and actions. That is, all nations and groups on the planet should be a full participating partner in the process. The membership of the regime should be as inclusive as possible. The process should give fair and equal opportunities for every party to voice concerns and views. In addition, the decision-making and deliberative process should conform to standards of procedure justice. It means the procedures should be transparent, issues should be addressed openly and fairly, and based on the latest scientific facts, among other things. There should be no hidden agendas that serve specific interest groups or blocs at the expense of the common interests. Only when processes and procedures comply with these conditions would it gain credibility and trust from participating members. Without credibility and trust, there will be no genuine commitment and actions from parties.

Moral Leadership

Nations most able to provide solutions to the problem should play leadership roles in the process. Nations that have made or are making most of the climatic impact should also act as leaders. But to sustain real success in responding to climate challenges, a new form of leadership – moral leadership is needed. In today's international negotiations or deal-makings, leadership is often driven by narrow national interests. Decisions are always dictated by short-term cost and benefit calculation. Economic and political as well as corporate interests invariably dominate the game. Worse still, many so-called "national interests" are in reality powerful special interests in disguise. Many nations are hijacked by special interests which corrupt or distort the true interests of their citizens. As said earlier, current political institutions, including democratic regimes may not be effective in designing and implementing climatic policies and actions that serve the global interest. Notwithstanding its strengths in other areas (e.g., in preventing tyranny by its checks and balances mechanisms), a democratic regime with its conventional leadership is subject to the electoral cycles of power shift, that are susceptible to short-term visions and solutions. As well as cantering to short term returns, *bona fide* national interests usually trump global interests when the two are in collusion. To counteract these problems, moral leadership as an alternative form of leadership is more needed to the climate regime.

Moral leadership (Cuilla, 1998, 2007; Rhode, 2006; Trevino and Brown, 2007) is characterized by its genuine concern for the common good, and the long-term well-being of humanity. Committed to right and justified values, moral leaders have the capacity (disposition, sentiments, intelligence) and intention to do the right things and to follow and develop ethical norms. They have the moral character to sustain moral acts and interactions with others with genuine concern for the well-being of people. They care about morality of means as well as that of ends, and are willing to guide or restrain narrow interests under the constraint of the common good and fairness. Recognizing messiness of world politics, they are no starry-eyed idealist, but have a pragmatic sense of making things work. They are also willing to make reasonable and tough trade-offs by balancing national and global interests, long-term interests and short-term interests through reason, mutual adjustments and benign compromises. Moral leaders need not be represented by a few powerful national leaders, though they indeed can play critical roles in shaping decisions. Their membership may not even be limited to top political leaders. Middle level officials or managers in governments, leaders and administrators in NGOs and industrial bodies can assume this role. In other words, leaders at various levels of government organizations or private institutions can be moral leaders. The more moral leaders participating in the negotiation process, the better will be chance of success.

Together these principles help define the basic structure of an ethical governance regime. They furnish the ethical basis for guiding and formulating strategies, norms, policies and measures, setting targets and schedules, and forming mechanisms and institutions for action, among others. The non-harming principle which stands as the supreme principle of ethics is assumed and not listed in this regime because it so basic and self-evident to be explicitly stated. Also, the moral imperative to act is so obvious that it is also presupposed in regime as well. That is, the ethical regime precludes non-action, or business-as-usual as an option.

Conclusion

What would an ethics-sensitive climate agreement that goes beyond the Copenhagen Accord look like? We do not have definite answers to this question. At the time of this writing, the Climate Accord produced in Copenhagen was disappointing and a failure in the eyes of many. In view of such a failure and disappointment, humanity needs more than ever more genuine cooperation to combat the climate change crisis. However, no cooperation that violates ethics is sustainable. Climate change has posed an unprecedented challenge to human society that requires intelligent, effective, and responsible collective responses. This surely is the defining moment for humanity. These responses are not possible without mutual trust and commitment. Ethics help build trust and commitment. Trust and commitment are critically contingent on the participant's perception on whether the cooperation is fair and equal, whether the collective undertaking conforms to the basic standards of ethics. Past experiences in large scale collaborative activities have shown that fairness and other elements of ethics secure and enhance stable cooperation, which are built on trust and commitment. Furthermore, legitimacy of the regime and its elements depend on ethics. Legitimacy is also a reliable predictor of why people are willing to follow norms. Neglecting or downgrading ethics in climate negotiation will have dire consequences. It will harm people's sense of justice, hurting their mutual trust and reducing their commitment. These together are detrimental to genuine participation and durable cooperation. Thus, durable cooperation requires ethics! An ethical climate regime intends to institute this critical element of success to the process, where it ought to belong. Through an ethical regime, norms to bind behaviors are not just treaty norms, but also ethical norms; responsibilities for actions are not simply treaty responsibilities but also ethical responsibilities. They are crucial elements to bind and guide behaviors and help raise the probability of getting the intended results, when working in tandem with other effective mechanisms. Surely, an ethical regime alone is not sufficient to bring about success. Neither does it underestimate the cruel reality of real-world politics. Other major elements, including good politics, objective and timely knowledge, finance, and technology, effective compliance, among other things, are also critical to the success. But ethics matter!

Postscript

The United Nations climate change conference held in Copenhagen December, 2009 produced a non-binding accord that was disappointing and ineffective. The accord was widely seen as a statement of intention which lacks concrete steps and specific time tables on action to combat the threat of global warming (Revkin and Broder, 2009). Items considered critical, including the GHG emission reduction targets for the mid or long-terms, were not included in the agreement. Obviously, it was the result of a reluctant compromise among the major blocs of powerful interests in the tortuous and flawed process of negotiations. For some, this failure may be indicative of the futility of the process framed by the United Nations Framework Convention on Climate Change and its related conventions and protocols which have guided the climate talks for over two decades. For others, flawed though it is, the Copenhagen accord is still a step forward in the right direction, preparing a stage for more substantive work to be completed. Perhaps a lesson learnt from Copenhagen is that the nations need a new way to tackle the problems. Some have suggested that instead of involving over 190 nations in discussions of board issues, a smaller bloc of 30 nations responsible for 90 percent of the emission will first negotiate under narrow agendas, including climate abatement financing, technology transfer etc. to come up with more workable solutions for a wider forum to consider and endorse. Whatever way the ensuing process will take, to obtain fair and credible solutions or proposals that nations will support, it seems that the climate change regime should satisfy the conditions that this paper has spelt out.

*A portion of this paper was presented at the Symposium on Sustainable Development and Quality of Life, at the National Central University, Taiwan, Nov 20-21, 2009.

References

Aldy, J. E. and R. N. Stavins: 2008, 'Climate Policy Architecture for The Post-Kyoto World', *Environment* 50, 6-17.

Broome J.: 2008, 'The Ethics of Climate Change', *Scientific American, June,* 97-102.

Brown D. et al.: 2008, White Paper on the Ethical Dimension of Climate Change, Rock Ethics Institute, available at http://rockethics.psu.edu/climate, accessed on October 2, 2008.

Chasek, P., D. L. Downie, and J. W. Brown: 2006: *Global Environmental Politics*, Fourth Ed. (Westview Press, Boulder, Colorado).

Ciulla J. B.: 1998, *Ethics, The Heart Of Leadership* (Praeger Publishers, Westport, CT).

Ciulla J. B.: 2007, 'What Is Good Leadership', in Ciulla, J. B., C. Martin, and R. C. Solomon, *Honest Work – A Business Ethics Reader* (Oxford University Press, Oxford), pp. 533-538.

Dietz, T., E. Ostrom, and P. C. Stern.: 2003, 'The Struggle to Govern the Commons', *Science* 302, pp. 1907-1912.

Gardiner, S. M.: 2004, 'Ethics And Global Climate Change', *Ethics* 114, 555-597.

Ip, P.K.: 2004, 'Corporate Responsibility and the Ethics Of Sustainability', *Journal of Humanity* 29, 69-92 (in Chinese).

IPCC (Intergovernmental Panel on Climate Change): 2007, *The Fourth Assessment Report*, available at www.ipcc.ch, accessed on March 20, 2008.

Ostrom, E.: 2009, 'A General Framework for Analyzing Sustainability of Social-Ecological Systems', *Science* 325, pp. 419-422.

Raymond, L.: 2008, 'Allocating The Global Commons: Theory And Practice', in Vanderheiden, S., ed. *Political Theory And Global Climate Change* (MIT Press, Cambridge, Mass.), pp. 65-86.

Revkin, A. C. and J. M.. Broder, 2009, "A Grudging Accord in Climate Talks", *New York Times*, December 20, 2009, http://www.nytimes.com/2009/12/20/science/earth/20accord.html?hpw=&pagewanted=print

Rhode, D. L.: 2006, *Moral Leadership – The Theory and Practice Of Power, Judgment, And Policy* (Jossey-Bass, San Francisco).

Shue, H.: 2001, 'Climate', in Jamieson D. ed., *A Companion to Environmental Philosophy* (Blackwell, Oxford), pp.449-459.

Singer, P.: 2002, *One Earth – The Ethics of Globalisation* (Yale University Press, New Haven).

Stern, N. et al.: 2007, *Stern Review on the Economics of Climate Change* (Cambridge University Press, Cambridge). Also available at www.hm-treasury.gov.uk/independent_reviews/stern_review_economics_climate_change/stern_review_report.cfm, accessed on April 6, 2008.

Turner, R.K. and D. Pearce: 1993, 'Sustainable Economic Development: Economic And Ethical Principles', in Barbier, E. ed. *Economics and Ecology: New Frontier and Sustainable Development*, (Chapman &Hall, London).

Trevino, L.K. and M.E. Brown: 2007, 'Ethical Leadership: A Developing Construct', in Nelson, D.L., and C.L. Cooper, eds., *Positive Organizational Behavior* (Sage, London), pp. 101-116.

Vanderheiden, S.: 2008, 'Climate Change, Environmental Rights, And Emission Shares', in Vanderheiden, S., Ed. *Political Theory And Global Climate Change* (MIT Press, Cambridge, Mass.), pp. 105-128.

CHAPTER TEN

Climate Change and Obligations to the Future

William Grey

Climate change is perhaps the biggest and most dangerous challenge that humanity has ever faced. But there are peculiar features of the problem which make it very difficult for us to address, and which taken together exacerbate the problem. One difficulty is that the magnitude of the problem means that dramatic and far-reaching changes need to be made quickly if we are going to avoid a very significant rise in global temperature during the course of the 21st century. Climate change is a complex problem which must be addressed by the physical and life sciences, psychology, economics and politics. But climate change also raises fundamental questions about our values, and it is therefore also an ethical question; indeed it is one of the most central and important questions in environmental ethics. Climate change raises very sharply the question of what sort of life we want to lead and the sort of life we want our descendants to be able to lead. Addressing the issues associated with climate change therefore involves grappling with a lot of empirical issues, as well as (among other things) metaphysical concerns about the status of future individuals, and ethical issues about the effects which our choices can have on biodiversity and the integrity of natural systems. These are difficult problems.

Ice core measurements show that the concentration of atmospheric CO_2 has varied between about 180 and 280 parts per million (ppm) over the last 600,000 years and that this concentration has been coupled with ice ages and interglacial periods, with the higher concentrations associated with the warmer periods. The concentration of CO_2 remained at about 280 ppm from the end of the last Ice Age, about 12,000 years ago, until the late 18th century, when industrial civilization began burning fossil carbon in the form of coal, and later oil. Over the last 200 years the concentration of CO_2 is increased by over 35% to its present level of more than 380 ppm, a concentration which has not been exceeded for the last 600,000 years and probably not for the last 20 million years. This significant increase is anthropogenic (human-caused) and has been caused mainly by burning fossil fuels and land clearing. Humans currently burn about 7 gigatonnes (7 thousand million tonnes) of fossil carbon per year, a quantity which significantly exceeds the planet's carbon sink capacity. There is an overwhelming consensus within the scientific community that anthropogenic greenhouse gases pose a serious problem. We are conducting a massive experiment in planetary engineering which is likely to see a significant rise in global temperature over the next century. Recent research at the Massachusetts Institute of Technology suggests that the predicted 4°C temperature rise by the end of the century may need to be revised up to as much as 9°C (Krugman 2009). A temperature rise of this magnitude and this rapidity would be catastrophic. However because global climate is a large and complex system, the projections about the effects of adding greenhouse gases involve uncertainty. There may be negative feedback loops which moderate changes as well as positive feedback loops which exacerbate them. We can nevertheless say that the overwhelming consensus, at least within peer reviewed climate science, is that anthropogenic global warming is real, and that it constitutes a serious problem, though there uncertainty and disagreement about how serious the problem is. The peer reviewed scientific consensus has been formulated by the Intergovernmental Panel on Climate Change (IPCC), established in 1988 by the World Meteorological Association (WMA) and the United Nations Environment Program (UNEP) to assess "the science, the impacts, and the economics of—and the options for mitigating and/or adapting to—climate change" (IPCC 2001: vii). The IPCC has produced four reports, the latest in 2007, which provide an authoritative guide about the state and direction of the problem.

Despite the overwhelming consensus of opinion among climate scientists, there are still some climate change deniers, some even with scientific qualifications—though usually not in climate science. Scepticism and dissent are important in scientific inquiry, because they help to ensure the integrity and accuracy of science.

So it might be argued that this vocal minority plays a useful role. Unfortunately they also generate public confusion. Nevertheless the peer reviewed science delivers a clear message. The 19th century German philosopher Schopenhauer said that all truth passes through three stages: first it is ridiculed, second it is violently opposed, and finally it is accepted as self-evident. The truth about climate change has emerged comparatively quickly in the last few decades and appears to be in all three stages simultaneously.

We have benefited greatly, and suffered greatly, through the use of fossil fuels. As George Monbiot has said:

> Fossil fuels helped us to fight wars of a horror never contemplated before, but they also reduced the need for war. For the first time in human history—indeed for the first time in biological history—there was a surplus of available energy. We could survive without having to fight someone for the resources we needed. Our freedoms, our comforts, our prosperity are all the products of fossil carbon, whose combustion creates the gas carbon dioxide, which is primarily responsible for global warming. Ours are the most fortunate generations that have ever lived. Ours might also be the most fortunate generations that ever will. We inhabit the brief historical interlude between ecological constraint and ecological catastrophe. (Monbiot 2006, xxi)

Climate change is generating a slowly emerging catastrophe. Unfortunately for our present predicament humans have inherited a brain which is very good at responding quickly to immediate threats, but which is very poor at detecting threats which emerge more gradually but which may have a much greater impact. (This is well illustrated in a number of the case studies discussed in Diamond 2005.) We are excellently adapted for risk assessment for the threats which confronted our Pleistocene ancestors, but poorly programmed to detect the risks which confront the civilized world of the modern industrial state in the 21st century. "In effect, evolution has programmed us to be alert for snakes and enemies with clubs, but we aren't well prepared to respond to dangers that require forethought" (Kristof 2009). Kristof illustrates this nicely by contrasting public response to two contemporary issues:

1. President Obama proposes moving some inmates from Guantánamo Bay, Cuba, to supermax prisons from which no one has ever escaped. This is the "enemy with club" threat that we have evolved to be alert to, so Democrats and Republicans alike erupt in outrage and kill the plan.

2. The climate warms, ice sheets melt and seas rise. The House scrounges a narrow majority to pass a feeble cap-and-trade system, but Senate passage is uncertain. The issue is complex, full of trade-offs and more cerebral than visceral—and so it doesn't activate our warning systems. "What's important is the threats that were dominant in our evolutionary history," notes Daniel Gilbert, a professor of psychology at Harvard University. In contrast, he says, the kinds of dangers that are most serious today—such as climate change—sneak in under the brain's radar.

Professor Gilbert argues that the threats that get our attention tend to have four features. First, they are personalized and intentional. The human brain is highly evolved for social behavior ("that's why we see faces in clouds, not clouds in faces," says ... Gilbert), and, like gazelles, we are instinctively and obsessively on the lookout for predators and enemies.

Second, we respond to threats that we deem disgusting or immoral—characteristics more associated with sex, betrayal or spoiled food than with atmospheric chemistry.

"That's why people are incensed about flag burning, or about what kind of sex people have in private, even though that doesn't really affect the rest of us," Professor Gilbert said. "Yet where we have a real threat to our well-being, like global warming, it doesn't ring alarm bells."

Third, threats get our attention when they are imminent, while our brain circuitry is often cavalier about the future. That's why we are so bad at saving for retirement. Economists tear their hair out at a puzzlingly irrational behavior called hyperbolic discounting: people's preference for money now rather than much larger payments later.

Fourth, we're far more sensitive to changes that are instantaneous than those that are gradual. We yawn at a slow melting of the glaciers, while if they shrank overnight we might take to the streets. (Kristof 2009)

Part of the problem, then, is our incapacity to detect slowly emerging threats. It is also true that the ethical framework which guides and sanctions our choices is poorly structured for long-term thinking. Our moral notions, such as duty and justice, focus on the rights, entitlements and obligations which we owe to our contemporaries, but do not adequately deal with intergenerational justice—the duties that we owe to our children and our grandchildren. Indeed some have expressed skepticism about whether we owe posterity anything at all. ("What has posterity ever done for us?" it has been asked.)

Stephen Gardiner (2006) has argued that peculiar features of the climate change problem present obstacles to making the hard choices which are necessary. Gardiner thinks that there is a convergence of a set of global, intergenerational and theoretical problems which constitute what he calls a "perfect moral storm". Three characteristics of the climate change problem which Gardiner identifies are first the dispersion of cause and effect, second the fragmentation of agency, and third institutional inadequacy. Moreover these characteristics each have a spatial and a temporal manifestation. Spatial dispersion means that the effect of any particular emission of a greenhouse gas is not realized at its source—its impact is global. Fragmentation means that climate change is not caused by a single agent, or even a small number of agents, but rather by a vast number of individuals and institutions each acting independently. While our collective impact may be huge, each individual act contributes infinitesimally to the problem. Institutional inadequacy for addressing the problem is a consequence of its globally pervasive nature—there is no adequate system of global governance to address climate change, and it is proving very difficult to develop an effective collective response to the challenge that it presents.

Greenhouse gas emissions have the structure of what Garrett Hardin called the tragedy of the commons (Hardin 1968). Each individual greenhouse gas emitter accrues a positive benefit from burning fossil fuel while the cost of the emission is fragmented and dispersed, both spatially and temporally. Under favorable conditions commons problems can be resolved through an agreement to change the existing incentive structure through the introduction of a system of enforceable sanctions (Gardiner 2006: 400). Hardin calls this solution "mutual coercion, mutually agreed upon". This forecloses the option of individual free riding, and collective rational action becomes individually rational also. However a problem we confront is the absence of mechanisms for globally enforceable sanctions—or at least a reluctance to deploy them.

Another source of reluctance to seriously address the problems is that the sources of greenhouse gas emissions are part of the wealth-generating structures of first world economies, and any drastic measures to curtail emissions may have significant impact on economic and political structures. Of course a sea level rise of between four and 20 meters will have its own impact on economic and political structures, but that won't be a problem for us—that will be a legacy that our great-grandchildren may have to deal with.

There may be a very significant sea level rise during the 21st century. Barring a mega project to cool the planet, the IPCC 2007 estimate of a 0.2 to 0.6 m sea level rise by 2100 may need to be revised upwards to at least 0.5 to 1.4 meter rise (most likely in the 1-2 m range), with a lag of several thousand years to reach an equilibrium 10 to 25 m higher than today (Ananthaswamy 2009). Admittedly these estimates are uncertain. But it would be extremely foolhardy to wait 20 years for better data in order to be able to announce with greater confidence that the sea level rise will be 2 m rather than the 50 cm estimated earlier. A rise of 1 to 2 m would be catastrophic, inundating many densely populated coastal regions including the Netherlands, Bangladesh, and Florida, as well as many of the world's major cities including Shanghai, London, Tokyo, New York and Melbourne. Urban planners are already restricting building in coastal danger zones. Historically, sea level has been up to 120 m lower during ice ages, and up to 70 m higher when the planet has been ice free during hot periods. But this is not relevant to us—we were not around at the time.

The three characteristics of the climate change problem which Gardiner identified—dispersion of cause and effect, fragmentation, and institutional inadequacy—are extremely problematic when considered from a temporal perspective. Gardiner calls this "the intergenerational storm". There is a significant temporal delay between cause-and-effect; the effects of warming, such as the melting of ice, might take a very long time to be fully realized. Also worrying is the fact that once emitted molecules of carbon dioxide can remain a very long time in the upper atmosphere. The IPCC says that the average time a molecule of carbon dioxide remains in the upper atmosphere is between five and 200 years. David Archer suggests that natural sequestration of anthropogenic CO_2 will take tens of thousands of years (Gardiner 2006: 403). The diachronic tragedy of the commons is more vexing than its synchronic manifestation. The problem with a temporally extend commons is that the parties who benefit and those who bear the costs of choices do not coexist, and this gives rise to a number of difficulties.

One of these problems which I will not address here in any detail, but which I have addressed elsewhere (Grey 1996), is what Derek Parfit has called the "non-identity problem" (Parfit 1984). Given some reasonable assumptions, it is plausible to suppose that *which* individuals come to exist is sensitive to, and will be changed by, the social policies which we adopt. So if we continue with 'business as usual' and pass on a ruined and degraded planet as our legacy, then our descendants who inherit that impoverished planet will have nothing to complain about since, if had we chosen to act otherwise, then they would not have existed at all, and their wretched and impoverished life is better (we may suppose they will suppose) than having no life at all.

But this cannot be right. For example, it is very likely that without World War II and Adolf Hitler my parents would never have met and I would not exist. However I do not have to be grateful for World War II and Adolf Hitler. I can be grateful that p, recognize that q is a necessary condition for p, but regret that q, which of course involves regretting that q is a necessary condition for p.

The world we live in is an impoverished world which would be richer if it contained dodos, passenger pigeons and thylacines, all now extinct. This is a fact which I regret. I have been harmed by my ancestors whose careless and insensitive environmental choices led to the extinction of thylacines, even though at the time of their bad choices I did not exist to be injured, and even though had they introduced policies and practices which supported the preservation of biodiversity I may never have existed. No doubt my 19th century Australian ancestors considered thylacines to be pests, to be exterminated in favor of European sheep and cattle. This was part of the so-called acclimatization movement's misguided ambition to Europeanize the Australian landscape. Our environmental sensibilities have improved since then, an example of what Dale Jamieson has called "morality's progress" (Jamieson 2002). The floral and faunal insults visited on Australia are now recognized widely, though unfortunately not universally.

Intergenerational obligations do not require us to be able to specify the identity of the individuals who will be the beneficiaries of our beneficent choices, or who will suffer the consequences of our poor ones. Such obligations need to be framed as what Parfit has called impersonal principles—principles which are such that we do not need to designate which persons whose rights we are concerned to respect, or who will be harmed by our failure to act with consideration. Parfit argues that this leads to an unacceptable consequence, which he calls "the repugnant conclusion". This is a problem which I think can be resolved, though I won't address it here. William FitzPatrick argues that even ordinary duties of care in familiar and straightforward cases need to be construed generally; that is, even in cases where the identity of the individuals liable to benefit from our care, or to suffer the lack thereof, is unproblematic. FitzPatrick (2007: 384) argues, convincingly in my view, for two important facts about rights—supposing that you want to use rights discourse to articulate our intuitions about obligations to the future.

> 1. It is possible to violate people's rights by *harming* them even if doing so *does not make them worse off* than they would have been had we respected those rights; and

> 2. An action can turn out to have violated a person's rights *without* its having been the case *at the time of acting* that the was any right constraining the agent from so acting.

That is, moral and political community extends beyond the boundaries of our contemporaries.

Another serious issue which arises in connection with considering the entitlements of future generations concerns discounting. Discounting may be reasonable for some issues concerning economic goods and investments. But there is no reason to suppose that nonmonetary goods can be discounted in the same way. Human welfare and well-being are important irrespective of temporal location. Parfit argues that discounting welfare itself quickly leads to absurdity (Parfit 1984: 484-5; 482). There might however be other reasons for discounting the future. Passmore (1974) suggests that while we may recognize claims for immediate posterity, there is too much uncertainty about the welfare needs of distant human generations to include them in our deliberations. This is unconvincing at least as far as it concerns a healthy environment in general and in regard to climate change in particular. The issue here isn't whether a future generation will be marginally better or worse off than we are. What is it at issue is preventing the possibility of people being much worse off than we are as a result of major catastrophe. James Lovelock believes us that the human population is much too large to be sustainable except in the very short-term (Lovelock 2006; 2009), a Gaian affliction for which he coins the label "polyanthroponaemia". Lovelock writes:

> Whatever our faults, we surely have enlightened Gaia's seniority by letting her see herself from space as a whole planet while she was still beautiful. Unfortunately, we are a species with schizoid tendencies, and like an old lady who has to share her house with a growing and destructive group of teenagers, Gaia may grow angry, and if they do not mend their ways she will evict them. (Lovelock 2006: 47)

There is a striking polarity of opinion about the human future between Cornucopian optimists and Malthusian pessimists. The former believe that we are on the threshold of unprecedented prosperity, while the latter believe we are confronted with major catastrophe. They might both be wrong, but they can't both be right.

Lovelock is the inventor and champion of the Gaia hypothesis, according to which living systems act in a collective and coordinated fashion to sustain optimal conditions for their own survival. On this view the physical conditions of the planet, such as temperature, atmospheric composition, and ocean salinity, don't merely shape the complement of life forms on the planet; this complement of life forms shapes those conditions through a variety of negative feedback mechanisms. Lovelock argues that we need not be concerned about the fate of Gaia. Life on earth will continue on for perhaps for another billion years despite our most destructive insults and predations. Eventually, though, increasing solar flux will overwhelm Gaia's temperature control mechanisms, transforming the planet into a barren desert. But if we act wisely we can sustain congenial conditions for a geologically long time—or rather, we can operate within sustainable limits and allow Gaia to maintain those conditions for us.

However while the life supporting planet will continue for hundreds of millions of years into the future it may not, if we are incautious, provide congenial conditions for civilized human life. We may for example reach a tipping point beyond which much of our life-supporting agriculture will collapse. A major challenge in the years ahead, if we are to maintain the sort of mild interglacial planetary conditions under which human civilization was established, is to work out how to live and flourish with carbon-neutral technologies—perhaps even carbon negative technologies, if that is what is required to respect the capacities of carbon sinks. This is certainly a challenge for agriculture (as well as for transport, running factories and for heating and cooling our buildings) but it is a challenge which, given political will (fortunately a renewable resource) we are able to address.

We should work systematically to sustain the sort of planetary conditions which have shaped us and to which we are adapted; that is a rich, biodiverse and relatively mild temperatured planet. Geological timescales provide a majestic conception of our ephemeral tenure as a part of Gaia. But they provide an irrelevant perspective for shaping the choices which we need to make right now which are relevant for human well-being and the continuation of civilized human life.

Over the last 3 billion years Gaia has shown many faces, many of them darker and less congenial for the flourishing of the likes of us–indeed most of them would have excluded us altogether. There was no atmospheric oxygen for Gaia's first 2 billion years, and by the time multicellular life finally emerged, about 500 million years ago, Gaia was already middle-aged. We should do whatever we can to sustain the present face of Gaia, either because the current assemblages of life forms are a magnificent legacy of billions of years of evolution, and should be preserved because of their intrinsic interest and value, or otherwise for the self-interested reason of ensuring our own survival. But it is very likely that we have already set in train an inexorable cascade of disturbances which, in the long-run, will be to the detriment of humanity. So some adaptation to anthropogenic change will almost certainly be required.

We must also do what we can to mitigate our impact on the planet by establishing, as quickly as possible, carbon-neutral technologies to secure a sustainable—or at least less unsustainable—future. We also need to extend our moral horizons to incorporate a more temporally extended sense of obligation if we are to achieve long-term sustainability. The moral heuristic known as the Golden Rule ("treat others as you have them treat you") articulates a parity of perspective—I understand and accept that your circumstances, which are relevantly similar to mine, warrant respectful consideration on my part, with the expectation that you will act with similar consideration and respect towards me. This is an important moral principle, and variants of it can be found in many cultures. We need to develop a temporally extended version of the Golden rule, very roughly "do unto future others as you would have had your ancestral others do unto you". Clearly extending our moral thinking to incorporate are much more extensive temporal perspective is going to require a great deal of work. That is one of the many tasks which environmental ethics will need to address.

References

Ananthaswamy, Anil. 2009. 'Sea level rise: It's worse than we thought', *New Scientist,* 1 July, http://www.newscientist.com/article/mg20327151.300-sea-level-rise-its-worse-than-we-thought.html

Diamond, Jared M. 2005. *Collapse: How Societies Choose to Fail or Succeed.* New York, Viking.

FitzPatrick, William J. 2007. 'Climate Change and the Rights of Future Generations: Social Justice Beyond Mutual Advantage'. *Environmental Ethics* 29: 369-388.

Gardiner, Stephen M. 2006. 'A Perfect Moral Storm'. *Environmental Values* 15: 397-413

Grey, William. 1996. 'Possible Persons and the Problems of Posterity.' *Environmental Values* 5: 161-179.
Hardin, Garrett. 1968. 'The Tragedy of the Commons.' *Science* 162: 1243-1248.

IPCC 2001. Third Assessment Report, *Climate Change 2001*, http://www.grida.no/publications/other/ipcc_tar/?src=/climate/ipcc_tar/

Jamieson, Dale. 2002. *Morality's Progress*. Oxford, Clarendon Press.

Kristof, Nicholas D. 2009. 'When Our Brains Short-Circuit', *New York Times*, July 2, http://www.nytimes.com/2009/07/02/opinion/02kristof.html?

Lovelock, James E. 2006. *The Revenge of Gaia*. London, Allen Lane.

Lovelock, James E. 2009. *The Vanishing Face of Gaia*. London, Allen Lane.

Monbiot, George. 2006. *Heat: How We Can Stop the Planet Burning*. London, Allen Lane.

Parfit, Derek. 1984. *Reasons and Persons*. Oxford, Clarendon Press.

Passmore, John. 1974. *Man's Responsibility for Nature*. London, Duckworth.

CHAPTER ELEVEN

Environmental Ethics in an Omniverse Environment: From Terrestrial Chauvinism to Golden Rule

Charles Tandy

§1 Introductory Remarks
§2 Initial Derivation of the Omniverse Model
§3 Fleshing Out the Omni Paradigm
§4 Temporal Reality, Including Time Travel
§5 Temporal Entities, Including S-Creatures and R-Beings
§6 R-Beings and Reason
§7 R-Beings and Knowledge
§8 From Terrestrial Chauvinism to Golden Rule
§9 Closing Remarks

§1 Introductory Remarks

An outline of reality, herein called the "omni" or "omniuniverse" or "omniverse" model, is presented and justified below. My self has a sense of <u>Personal</u> (or self) Reality that is influenced by <u>Temporal</u> (or contingent) Reality and by <u>Paragonal</u> (or necessary) Reality in an Omniverse Environment (the omniverse is all of reality). The paper discusses the nature and obligations of temporal personal entities with the ability to reason and be reasonable ("r-beings") in an omniverse environment. A. N. Whitehead and J. Bronowski prove helpful here. R-beings with the limited reason of humans have an obligation to become advanced r-beings, and advanced r-beings have an obligation to advance further and further. As r-beings advance, they outgrow the chauvinism of my-species and my-planet. With perpetually advancing knowledge gained from scientific method and golden rule, r-beings are able to improve world and self. With this in mind, the paper articulates ethical-political and other details or implications for r-beings in the historical position humans find themselves today.

THE OMNIVERSE

PARAGONAL Realm

TEMPORAL Realm

PERSONAL Realm
(intimations of:)

> Personal realm
> Temporal realm
> Paragonal realm
> Omniversal realm

> **THE DIAGRAM OF THE OMNIVERSE MODEL**
> of reality (on the previous page) is an EPISTEMOLOGICAL
> account that begins with My Self (in the center).

Some readers may more or less disagree with the proposed omniverse account of environment but yet find it possible to more or less agree with the indicated ethical shift "from terrestrial chauvinism to golden rule". On the other hand, the reverse may also be true. Some readers may more or less disagree with the indicated ethical shift "from terrestrial chauvinism to golden rule" but yet find it possible to more or less agree with the proposed omniverse account of environment. (Naturally, I prefer to think of the indicated ethical shift and the proposed omniverse account, as pointing to each other.)

§2 Initial Derivation of the Omniverse Model

I wanted my model to include all ("omni") of reality rather than part of reality – and my perspective on reality is that of a philosopher rather than a physicist. So I chose a term that should mean all of reality – not cosmos or universe or multi-verse, as a physicist might do, but omniverse (omni-universe). I will now proceed to derive or justify my omniverse model briefly as follows:

(A) The personal is real;
(B) The temporal is real;
(C) The paragonal is real;
(D) The omniverse is real;
(E) The omni model is unreal;
(F) The omni model is relevant.

(A) The personal is real. Reminiscent of Descartes, I begin construction of the omniverse model by showing to myself that I am real. (You may be able to apply this reasoning to show to yourself that you are real.) To wit: I am aware that I am reasoning; (therefore,) I am aware that I am; (therefore,) I am aware; (therefore,) I am: Therefore→ •I am; •I am aware; •I am aware that I am; •I am aware that I am reasoning. (Accordingly: The personal is real.)

(B) The temporal is real. If it is the case that I am no longer aware (possible examples: dreamless sleep; death), it will nevertheless always be the case that I <u>was</u> aware. My reality (and perhaps your reality) has temporal (contingent) aspects to it (perhaps including experiential blanks). Whether I like it or not, I am a self undergoing experiences in time. (Accordingly: The temporal is real.)

(C) The paragonal is real. It is necessarily the case that $1 + 1 = 2$. There are paragonal realities we associate with such mathematical and logical necessities. There are also paragonal realities we associate with ethical values. Event A or decision B may be in my or our objective ("best") interest. Alternatively, event A or decision B may be objectively ("really") harmful to me or us. Such is objective ethical reality even if we are not always certain about the objective ethical status of event A or decision B. Likewise, we may not be very knowledgeable of mathematical realities. Nevertheless, it seems that mathematical and ethical realities are necessarily the case. (Accordingly: The paragonal is real.)

(D) The omniverse is real. As previously indicated, by omni or omniverse (omni-universe), I mean all of reality. Although all of reality (the omniverse) is necessarily a unique concept-reality unlike any whole or universe or other reality within the omniverse, it seems fair to say that all of reality is real. (Accordingly: The omniverse is real.)

(E) The omni model is unreal. I have said that the "concept-reality" omniverse is unique. Nevertheless it must be pointed out that concepts and models (paradigms or theories), as such, are not real (other than being concepts/models). My omniverse paradigm is not the omniverse! At any moment the omniverse may well act to upset my simulation of it. (Accordingly: The omni model is unreal.)

(F) The omni model is relevant. Perhaps you prefer reality to paradigms or models? Unfortunately you do not have much choice in the matter. Let me explain. You interpret the messages you receive as helpful or hurtful based in your favorite paradigm or value-system. Yet a message or its interpretation is open to question. Like it or not, you sometimes receive messages which are illusory or misleading. Thus, if your old paradigm doesn't seem to work, you may search for a new one. If you lack (much of) a paradigm or model (whether old or new), then you have little or no knowledge of reality. Kenneth Boulding (1956) has pointed out that there is a sense in which "there are no such things as 'facts.' There are only messages filtered through a changeable value system." (p.14) And, as "Francis Bacon wisely observed in his *New Method*, 'truth will sooner come out from error than from confusion.'" (Barzun and Graff, 1985: p. 426) In other words, living the life of an ostrich is not an idea to be seriously entertained. (Accordingly: The omni model is relevant.)

§3 Fleshing Out the Omni Paradigm

Of what general kinds of reality are there? There are necessary kinds of reality and there are contingent kinds of reality. Contingent (or temporal) reality includes non-necessary entities such as numerous alternative universes with alternative (contingent) "universal laws". Necessary (or non-temporal) reality includes e.g. mathematical forms and ethical values. See §2 (C) above; also note that Rickert (1902) and Li (2002) offer two very different approaches to defending objective values (valid values or objective interests, respectively). I find each approach to defending ethical paragonals persuasive.

Thus the omniverse may well contain many different kinds of universes and many different kinds of beings. What we experience as things of a "physical" kind may be an experience not available in some other universes or not possible for some other beings. Likewise there may be a variety of kinds of experience or realms of reality not available in our particular universe and not possible for beings like us.

Yet there are all sorts of <u>apparent</u> "realities" that provide us with "impossible" experiences. This includes science fiction/fantasy movies, 3-D holographic effects, virtual reality machines, everyday common illusions as the "broken" stick in water, and the delightful tricks we happily experience at magic shows. Advanced beings could presumably not only engender universes with laws tailored to their specifications, and intervene contrary to those laws, but they could also use those laws and interventions to produce strongly convincing virtual realities or appearances in apparent contradiction to "natural laws" – and more, even "contradicting" the laws of mathematics and logic. This is perhaps just the sort of thing Grand Magicians or Advanced Beings or Magisters Ludi would enjoy doing. (Indeed, Descartes was famously concerned with the possible mischief of an "evil demon".)

One difficulty we have is our huge ignorance of the omniverse both temporal and non-temporal. Given such immense uncertainty, how could we ever know if we daily live our lives in a real or illusory world? Since "ever" is a very long time, perhaps we should enjoy the very long adventure. So let us ask: What ought our first step be in this long journey much longer than a thousand miles or a thousand years?

First of all, like Hume, I don't think it practically wise for us to extend Humean skepticism to our everyday lives. If we are in a game, we are not likely to win by not playing the game. Although it seems we are presently far from being Magisters Ludi or Advanced Beings or Grand Magicians of the required sort, still we are not totally ignorant of the game. We have some limited knowledge of the omniverse – and we can see ourselves becoming more and more knowledgeable over time.

On occasion we can use our limited knowledge of necessary realities such as mathematical forms to triangulate and identify mere appearances. The magician or the illusionist may seem to tell us that $1 + 1$ does not always equal 2. So when we find such an anomalous appearance of the grosser sort, we try to figure out the trick. We may conclude that the two raindrops, now become one, have a volume equal to the two raindrops. Or we add two units of liquid together and get something distinctly more or distinctly less than two units. In such case we may try to invent a new scientific theory to explain the results – we do <u>not</u> say that we have falsified the necessary reality $1 + 1 = 2$.

Consideration (1): With the help of necessary realities related to Gödel's logical proofs (or for other reasons, but Gödel will be explained in §7 below) we may conclude that it is certain or likely that the omniverse is infinite, that necessary realities are infinite, and that contingent realities are infinite. Presumably a large percentage (99.99% density of infinity?) of advanced beings would know this as well.

Consideration (2): With the help of necessary realities related to ethical values (or for other reasons) we may conclude that a large percentage (99.99% density of infinity?) of advanced beings would know of matters related to what we loosely call "the golden rule".

Consideration (3): If we put considerations 1 and 2 together, then we seem to get something analogous to "the veil of ignorance" (hypothetically) posited by John Rawls (1971) for the purpose of identifying proper or just political arrangements. But now we can see (given 1 and 2 above) that the (veil of) ignorance is real, not just hypothetical; it is real not only for us but for the advanced beings as well. In other words, there is now a real motivating force for "us" (whether human-beings, advanced-beings, advanced advanced-beings, et cetera ad infinitum) to behave toward each other in a "golden" way. (Perhaps we will never be asked to join "the Galactic Club" but rather will automatically become a member of "the Golden Club" when we learn to follow the "golden" way?)

What exactly this all means for our particular planet or our particular universe may not be altogether clear. As Amartya Sen (1999) and John Rawls (1999) have pointed out, though, a people's self-respect and self-development is vitally important to their developmental success. (This is a reason why some resource-rich countries ultimately fail while some resource-poor countries ultimately succeed.) Respect for one's own autonomy and respect for the autonomy of others seem to be related to the necessary realities I have called ethical paragonals or the golden way. How are the infinity (?) of "ethical values" in the realm of necessary reality related to each other? How do advanced beings proceed to attempt to weigh values (and conditions) so as to make optimal decisions with respect to a variety of kinds of universes and kinds of beings? (Alas, these relevant questions are somewhat beyond the scope of the present paper.)

§4   Temporal Reality, Including Time Travel

It seems that it is always reasonable to ask what, <u>if anything</u>, happened before and after event T (e.g., the big bang beginning of a universe). One may imagine the answer: "Nothing happened <u>except</u> (the "arrow" of) time: T-1, T-2, T-3, et cetera ad infinitum; T+1, T+2, T+3, et cetera ad infinitum. In this sense one may say that the <u>whole</u> temporal realm is infinite with respect to both the past "befores" and the future "afters" (perhaps this is related to the so-called "B theory" of time). On the other hand, at least in terms of the reality of human beings and/or advanced beings, we can say we have some limited free will and some limited ability to influence the details of events that take place <u>within</u> the temporal realm (perhaps this is related to the so-called "A theory" of time). Although deaths and other changes that take place within the temporal realm are important to us and are to some limited extent influenced by us, these deaths and changes should <u>not</u> be confused with the temporality of the temporal realm (the entire environment of infinite 'befores" and infinite "afters")! Charles Hartshorne (1951) chose to use the language of Whitehead's process philosophy to express this distinction, as follows (p. 542, emphasis in original):

> The later event prehends the earlier and so contains it, but the converse is not true; and this one-way relationship remains even when both earlier and later events are in the past ... no matter how fully their original immediacy is preserved. Obviously, it is not

because of fading or perishing that earlier is contained in later, though later is not contained in earlier. It is rather *in spite of perishing*. Were loss of immediacy the last word, how could the faded event in its non-faded vividness, as it was when present, be contained in the new present? Yet such containing is the theory of succession under discussion. It is the reality of the new *as added to that of the old*, rather than the unreality of the old, that constitutes process.

At any epoch in time (very short or indefinitely long), deaths may or may not occur – the temporal realm is, so to speak, indifferent to such details. And, as just explained, not only is the present real, but the past is real also. With respect to what has been called "practical time travel" – and matters related thereto – my previous analysis of the temporal realm (Tandy, 2006) draws the following six conclusions:

●1. The past exists as an expanding fixed unity.

●2. The present is the leading edge of the past as it expands.

●3. The future is not yet fully determined/fixed.

●4. The underdetermined future as it proceeds to become more nearly past (fixed) is influenced by the expanding fixed unity (the past), including by free agents of good will.

●5. Sooner or later, barring catastrophe, it seems highly likely that technology will advance so that the capacity for forward-directed time travel is possible. [Suspended-Animation (per molecular nanotechnology) and Superfast-Rocketry (per relativity physics) are examples of forward-directed time travel.]

●6. Sooner or later, barring catastrophe, it seems likely that technology will advance so that the capacity for past-directed time travel is possible. [Time-Viewing is one example of past-directed time travel.]

A more tentative <u>seventh</u> conclusion was that the concept of intrinsic time or intrinsic history (i.e., the intrinsic-temporality of the time-traveler, as distinguished from either merely-subjective time or literal-clock time) "is especially helpful in characterizing whether time travel did or did not occur in a particular circumstance." (pp. 383-384) If one travels backward in time in the ("many-worlds") omniverse, one does not come from the past but from the future (i.e., from the unique time or history intrinsic to the unique time-traveler). The temporal realm (the omniverse's temporal environment as such) has its own ("arrow" of) time, but it is another (different) matter that (in addition) each temporal entity <u>within</u> the temporal environment has its own unique intrinsic time (history). According to my proposed general-ontological schema, but unlike almost all physical-scientific theories of backward time travel, it would seem that in principle <u>any</u> past time and <u>any</u> universe is a candidate for visitation. (The ethics of time travel or inter-universe travel is another matter.) Moreover, the time-traveler – a temporal entity having its own unique intrinsic time <u>within</u> (and thus different from) the omniverse's temporal realm as such – may be an atom, a human, a planet, or a universe.

§5 Temporal Entities, Including S-Creatures and R-Beings

Above I have reasoned (or, like a good magician, waved my hands to show) that there are paragonal (non-temporal, necessary) aspects of reality and temporal (contingent) aspects to reality. Although other universes in the omniverse may differ, the following appears to be true of our universe or our little corner of our universe: Within the temporal realm of our tiny region, there are nonpersonal entities and personal entities, as follows: [1]

• <u>Temporal nonpersonal entities</u> include: Energy (Quanta); Matter (Atoms); and, Life (e.g., Flowers) (Biosystems).

- <u>Temporal personal entities</u> include: Sentience (e.g., Swans) (S-Creatures); and, Reason (R-Beings). Some r-beings are better at reason (reasoning or being reasonable), than others – to wit: Humble Reason (e.g., Human Beings) (H-Beings); and, Advanced Reason (Advanced Beings) (A-Beings).

Note that in our consideration of to what extent a particular temporal entity is (1) a nonpersonal entity; (2) an s-creature; or, (3) an r-being – we should obviously <u>not</u> base the evaluation on the <u>species</u> to which the being is said to belong. For example, some individual members of the human species (newborn humans; adult humans continuously severely mentally impaired from birth) do <u>not</u> belong in category (3) above. For example, some individual members of non-human species (some individual non-human animals) <u>do</u> belong in category (3) above. The "real-life" boundaries between the species are <u>not</u> sharp; in addition, the "real-life" boundaries between the three categories above are <u>not</u> sharp. The present "anti-speciesist" paragraph should be kept firmly in mind when correcting or correctly-interpreting the present paper.

One may also note that according to Confucius, *ren* ["1=" roughly approximates the Chinese symbol for *ren*] is necessary for true learning as distinguished from mere cleverness. *Ren* may be translated as benevolence or fellow-feeling. Thus perhaps it is wise to identify an r-being with an r-being of the *ren*-being kind.

§6 R-Beings and Reason

Whitehead's ***The Function of Reason*** (1929) explicitly specifies three desiderata if we are to function as reasonable beings ("r-beings", whether human beings or advanced beings); the three functions of any reasonable being are: [2]

1. Living or surviving (as distinguished from dying-to-death or extinguishing-to-extinction).
2. Living well (as distinguished from mere surviving).
3. Living better and better (as distinguished from just living well).

If we combine this with "golden rulish" (empathy/sympathy) considerations, then we can apply these reasonable functions or healthy motivations both to individual humans and to humankind (civilization). Thus: Human-beings should strive to become advanced-beings, advanced-beings should strive to become advanced advanced-beings, et cetera ad infinitum. Human civilization should strive to become trans-civilization, trans-civilization should strive to become trans trans-civilization, et cetera ad infinitum. Bostrom and Roache (2007) have emphasized the goal of individual survival for the purpose of becoming better than well. They have also emphasized the special importance of the survival of humankind; if humankind is extinguished, then no human individual will be able to live, to live well, or to become an advanced being.

Reasonable beings ("r-beings") have the ability to reason about the shared purpose of all r-beings, whether human or advanced. Advanced beings may be better reasoners than humans, but they both have, e.g., the capacity to respect each individual r-being and to respect all r-beings as a whole. Inspired by Whitehead's *The Function of Reason*, I will attempt to elaborate. [3]

Advanced or transhuman beings (or the so-called "Singularity") may not be altogether different from lesser r-beings, even those who are as severely challenged emotionally and intellectually as is the case with humans. The range of emotion and intellect is extremely narrow in human beings, but not so deficient as to altogether absolve them of ethical responsibility. A wider range of emotion and intellect means that advanced beings have a greater ethical responsibility than do lesser r-beings.

Within the temporal realm there appear to be two great contrasting tendencies. One is decay (degradation): Things fall apart or simplify. The other is evolution (renewal): Things become more complex or creative. Apart from input by r-beings, evolution is blind or indifferent or anarchic. Fortunately, the self-disciplined creative reason of r-beings is sometimes able to discipline evolution (regulate matters within the temporal environment) so as to make it sighted or purposeful or ethical. (R-beings exercise "moral dominion".)

Thus: A purpose of r-beings is the exercise of moral dominion and the promotion of diversity. R-beings are engaged in the art of life and living. Rocks or atoms are better at survival than are plants or animals. Yet r-beings consider survival of life important, and are not content just with rocks or atoms. R-beings don't merely live in an environment – they actively change or regulate their environment. For them the art of living involves not only survival, not only living well, but perpetual advancement or enhancement (i.e., living better and better).

Two major aspects of the ability of r-beings to engage in reasoning have long been abstracted (identified) by philosophers: "speculative" reason and "practical" reason. The first may be identified with the godlike wisdom (or complete understanding) sought by the philosopher Plato. The second with the foxy cleverness (or immediate method of action) portrayed in the fantasy-hero Ulysses.

If we are not careful, successful cleverness may convince us that Plato was a fool. Against such half-way cleverness a diversity of methods or approaches would seem wiser and may generally help guard against trained incompetence or a hegemonic methodology. We must be vigilant: The methodology of a special discipline (the self-discipline of a methodology) should never replace the self-disciplined creative reason of r-beings. The self-disciplined creative reason of r-beings signifies more than a (life of existentialist) rebellion against the absurd; the moral dominion of r-beings constitutes an actual counter-agency not only to hegemonic thinking but indeed to temporal decay. The active purpose of r-beings is to save and remake the temporal world. It is a perpetual striving toward the infinity we call the golden age or the golden rule or the infinite game. [4]

Practical reason, unlike speculative reason, is concerned with staying alive and with ethical behavior. But speculative reason has a disinterested curiosity that desires understanding even of all the omniverse; it assumes life as a given; it seeks better and better life. This better life is a process of betterment in the sense of better understanding for its own sake (disinterested curiosity about all things). "Throughout the generality of mankind it flickers with very feeble intensity." (p. 38) And it "is tinged with bitterness ... of an ultimate moral claim." (p. 39)

It is the advancement of mathematics and logic that gives method or discipline to speculation. Thus, instead of mere aphorisms and inspirations, we can produce a variety of systems of thought we call religions, philosophies, and sciences. Such religious, philosophic, and scientific systems must be perpetually open to modification if they are to progress in a reasonable way.

It is the interaction of the old reason (practical reason) and the new reason (speculative reason) that has given us modern science and modern science-based technology. Such interaction may historically soon give us a modern ethics and a modern ethics-based politics. But such advance may be resisted by obscurantism (the old insistence of practical reason that free speculation is dangerous). In any given historical epoch, obscurantism may be practiced by those dominant in religion, philosophy, and/or science.

Speculative reason (e.g., speculations by philosophers during the European Middle Ages) may build up a huge reservoir of apparently unfruitful concepts over many decades or centuries; then, with a little assistance from practical reason and the historically new environment in which the new scholars find themselves, suddenly there is a great breakthrough producing many fruitful results. Unaware that they would have failed without the huge reservoir of concepts built up by speculative reason, they think they are responsible for the "magical" results. Perhaps a bit too harshly, Whitehead explains practical reason's blindness to the major background cause (speculative reason) for its new success (modern science) – as follows:

> There is a large audience, a magician comes upon the stage, places a table in front of him, takes off his coat,

turns it inside out, shows himself to us, then commences voluble patter with elaborate gestures, and finally produces two rabbits from his hat. We are asked to believe that it was the patter that did it. (pp. 57-58)

Speculative reason seeks to understand all methods and to transcend all method with a higher, comprehensive understanding. [5] This quest for infinity is forever unattainable by r-beings. It is pursued for its own sake.

Speculative reason holds in trust for future generations its growing supply of creative concepts and disciplined constructions. Mathematics was a mere curiosity for many centuries – until mathematical physics appeared. "The ultimate moral claim that civilization lays upon its possessors" Whitehead advises, "is that they transmit, and add to, this reserve of potential development by which it has profited." (p. 72)

Practical reason can help us live and live well. But speculative reason not only helps us live well – it helps us live better and better. The objective of the discipline of speculative reason is not stability but betterment. Up to this point in history our ability to reason, with reference to both speculation and practice, has been dismal. "But it is there," observes Whitehead; r-beings already have some limited knowledge "of that counter-tendency which converts the decay of one order into the birth of its successor." (p. 90)

§7 R-Beings and Knowledge

Jacob Bronowski's ***The Identity of Man*** (1971) explicitly discusses the difference between "men" (selves/minds of some richness) and "machines"; I will use some of his thoughts to provide possible insight into the nature of reasonable beings ("r-beings", whether human beings or advanced beings). Inspired by Bronowski, I will now attempt to elaborate. [6]

We think of all r-beings as "one of us" – but yet each of us, each self (r-being) is, and wants to be, a free agent different from other r-beings. On the other hand, the nature of machines (as mere mechanisms or formalized operations or algorithms) is to be law-abiding. But r-beings have the capacity to break out of nature via free agency. "My way": An r-being wants to be free to be itself, to be different from others. An r-being may actively decide to behave differently when the same situation occurs a second time simply because it knows it is not the first time.

R-beings turn their growing experience into growing knowledge and their growing knowledge into a growing readiness for action (modification or betterment of self and environment). The r-being is not fixed, but is a process of unending growth. Much of this growth and growing experience happens or is produced inwardly rather than outwardly. The r-being's mind actively works with images and thus has an "imaginary" (fictional/non-existent?) life – recalling, fantasizing, speculating, foreseeing.

A machine has unambiguous input and unambiguous output. A respirator machine or one's mechanical non-conscious breathing is vitally important; we would die if we had to continuously decide whether to breathe or not! The importance of such unambiguous input and output of air should not be ignored in our analysis of life and world.

R-beings derive knowledge based on two modes of experience. (1) Some kinds of knowledge are formal (or can be formalized): I "hit" my fellow colleague at the symposium. (2) Some kinds of knowledge are informal (or can not be formalized): I "embarrass" my fellow colleague at the symposium. The informal kind of knowledge is self-knowledge: I recognize myself in my fellow r-being. But a machine does not recognize itself in an r-being way: "we cannot now conceive any kind of law or machine which could formalize the total modes of human understanding." (p. 25)

Although the infinity of all future physical science can never be formalized, at any given point in time our (incomplete) science of the workings of the physical world can be (tentatively) formalized. [7] All r-beings, as integral to practical action, form a picture of the world. This picture changes as their experiences grow.

In humans the pleasure and pain centers are found mostly in the (evolutionarily) older part of the brain. Many sensory and sensory-interpretation functions are performed either prior to reaching, or without ever reaching, the human's brain. In terms of capacity to engage in the formal procedures of classical logic and precise calculation, a human-being is far inferior to the machine-computer.

Often humans do not use such a logic of strict certainty (they lack such capacity except on a minor scale). So the human's brain attempts to construct a picture of the world rather than engage in precise calculation. The picture it constructs is not one of certainty and precision. Tentative, fallible decisions are made as to whether this is real or that is illusory. (In recent decades, philosophers have introduced the epistemological idea of "reflective equilibrium.")

These considerations suggest that the newer human brain is not about the precision of calculation so much as about the widening of consciousness. The images in the brain increase (widen); with this widening, the interaction between the brain and the senses widens. Thus both our physical (science) knowledge and our self (r-being) knowledge expand. Our knowledge widens while remaining tentative: "certain answers ironically are the wrong answers." (p. 41)

Thus physical-science is part discovery and part invention; it is a kind of language for describing the physical world. Our images or concepts are the <u>vocabulary</u>. The arrangement of the concepts into "laws of nature" is analogous to <u>grammar</u>. A <u>dictionary</u>-like translation of the grammar tells us the relevant observations to test. Language (and therefore science) is a perpetually living, open, changing process.

The question of whether we should arrange our concepts into grammar (laws) is less interesting than how we form such arrangements. Our science is based on disciplined guesses and generalizations. Our scientific laws are not forecasts but fallible, unifying explanations.

The imaginative <u>processes</u> <u>of</u> discovery differ from the formal (mathematical-logical) <u>display</u> <u>of</u> discoveries. At any given point in time, formal science thus displays itself as a closed system; but science as creative process is an open system. Science begins with imagination and then seeks to implode or minimize the ambiguities it finds. Poetry begins with imagination and then seeks to explode or exploit the ambiguities it finds. [8]

Science helps us gain knowledge of the physical world; the arts help us gain knowledge of the world of self (selves, r-beings). Although an r-being as a self sees itself uniquely from the inside, our science seeks to provide us all with knowledge of a common external world. By identifying yourself with other r-beings, you may not learn how to reason or how to act – but you may gain knowledge of yourself.

Science provides us with an "as if" final language, but at every stage the language of the arts is open. The arts cannot be understood unless we understand what it is to be an r-being (self). Both science and the arts begin in the imagination (mental images, not physical sensations). The arts enhance our experiences of being; science enhances our technologies of action.

R-beings have conscious imagination. Only an r-being has the ability to converse with itself. An r-being consciously knows it exists in an environment: "no other animal seems to be able to draw a clear boundary between himself and his environment. His memory is too short and his habits are too strong to make him firmly distinguish what he does from what is happening to him." (p .90)

There is me (myself) and there is not-me (environment). If the young r-being grows up in an environment devoid of other r-beings, then the youngster will have little understanding of what it is to be an r-being (knowledge of self will be extremely limited). Growing up in a culture (most any culture) is better than growing up devoid of culture or r-beings. Having a worldview (most any worldview) is better than not having a worldview. Cultures, worldviews, and r-beings change over time; not infrequently, change is from bad to better (instead of from bad to worse).

An r-being can recall what it no longer sees (perhaps a human gains this ability at about six months of age). For an animal: Out of sight, out of mind. For an r-being: Absence makes the heart grow fonder.

While all biology follows the arrow of time, r-beings are consciously aware of the future and consciously direct their actions with the future in mind. R-beings are aware that they are different from their environment; and R-beings are aware that they are beings living from the past into the future. The two halves of being an r-being are: imagining our future environments with the help of science; and, imagining our future selves with the help of the arts. [9]

The process called a self or r-being is not altogether a machine – since such a dynamic is not identical to any mechanism (algorithm or code of instructions). Rather, we experience and develop images in our minds. Some of our mental images name particulars, but mostly they name kinds (types). Every creative imagination, like every natural language, necessarily has ambiguities in it. Science uses concepts that apply not to the unique selves of r-beings but to a common physical world. Science seeks to reduce ambiguity. The arts use concepts relevant to the uniqueness of a self and seek to amplify our experiences; poetry has no need to resolve ambiguities because it is about empathy (being), not about provoking action to resolve differences.

In the past we have been more concerned that the members of our tribe have similar beliefs than that the beliefs be true. But in a scientific age of weapons of mass death and destruction we can no longer give truth a back seat. It is possible to create a common coherent philosophy and politics (and common task) for all r-beings that gives both truth and empathy a front seat. Empathy (or sympathy or the golden rule) involves self-respect and respect for all r-beings.

The search for truth about self and environment is a never-ending common duty: "this assumes that the truth has not already been found ... [and] that it is not there to be found, once for all." (p. 115) A society that has found the truth is an authoritarian culture. A truth-seeking society values originality, independence, and dissent. Justice and freedom are central to the protection of these truth-seeking values. Tolerance is based on respect for self and all r-beings.

The philosophy of r-beings is derived from physical knowledge (science) and empathetic knowledge (the arts). The two values, tolerance and respect, taken as one, we may call dignity. Dignity serves as an overlap or bridge between the puritan values of science and the intimate values of the arts. Dignity links society and individual; thus the r-being is "the unique and double creature: ... the social solitary." (p. 121)

The specialness of the r-being is based not on its experience of the physical world but on its experience of other r-beings. With scientific knowledge the r-being acts to become the master of creation. But: "The knowledge of self does not teach him to act but to be; ... it makes him one with all creatures." (p. 122)

Logical proofs provided by Kurt Gödel and Alfred Tarski in the 1930s have profound implications for our search for knowledge. [10] What is to be said about the symbols of "a formal logical language ... comes not from physics and chemistry and biology, but from symbolic logic." (p. 129) During the 1930s the logician Kurt Gödel proved that "a logical system which has any richness can never be complete, yet cannot be guaranteed to be consistent." (p. 130) Also during the 1930s the logician Alfred Tarski proved that "there can be no precise language which is universal." (p. 131) These "Gödel-Tarski" theorems, taken together, I will call "g-t theorems". According to Bronowski (p. 131):

any logical system to which they [the g-t theorems] apply must include the arithmetic of whole numbers as a basic part, and they must be distinguishable from the rest of the continuum of quantities. But with this proviso ... they apply to any system of thought which attempts to set up a basis of fundamental axioms and then to match the world by making deductions from them in an exact language – the language of physics, for example, or the chemical language inside the brain.

Thus the classical model (or arrogant ideal or noble dream?) of science is hopeless. Any set of scientific axioms is necessarily incomplete; and, necessarily, any set of scientific axioms must always be open to the possibility that it may be shown to be inconsistent. Any science seeking exactness inevitably and perpetually has these considerable limitations. "That is, only an axiom which introduces a contradiction can make a system complete, and in doing so makes it completely useless." (p.132) At a given point in time, scientific knowledge may seem to have achieved a universal, consistent, closed language. But more wisely, the actual language of scientific discovery is always open, not to be represented in the form of a logical machine.

Considerations above seem to insist that paragonal realities (e.g., the unlimited realms we identify with mathematics and logic) are infinite and that temporal realities (and the never-ending adventures of r-beings for physical-scientific knowledge) are infinite. It seems that the omniversal or cosmological default position should not be that there are many worlds but that there are an infinity of worlds and an infinity of r-beings. In principle the potential for an r-being to expand its consciousness, and to expand it again and again, is infinite. The omniverse is infinite in an infinity of paragonal, temporal, and personal dimensions.

"Any finite system of axioms can only be an approximation to the totality of natural laws. ... [Thus natural laws in] their inner formulation [paragonal reality?] must be of some kind quite different from any that we know." (p. 133) Our formal logic is not the logic of nature; from time to time our system of science must be enlarged. Scientific discovery lies outside our formal logic.

R-beings are imaginative and their imagination is free, beyond the bounds of formal logic. Unlike lesser beings, an r-being has the ability to refer to itself – accordingly, the natural language of r-beings is not a machine language. Philosophy requires natural language even when it finds machine language useful. Although logic-and-mathematics is often reliable, nevertheless it breaks down when it refers to itself. But philosophy has a more severe problem since self-reference is integral to it. The axiomatic system is only partly suited to predictive fields such as logic, mathematics, and physical science. The axiomatic system is even less suited to non-predictive fields such as philosophy, social science, and the arts and humanities. The self-knowledge we associate with r-beings "cannot be formalized because it cannot be closed, even provisionally; it is perpetually open, because the dilemma is perpetually unresolved." (p. 146)

The r-being is able to use its imagination to understand self-reference in a way not possible by an algorithm or machine. The imaginative logic or creative imagination of the r-being differs from the formal logic of the machine. R-beings engage in personal (self) reference and thus are able to identify with all r-beings (as in the golden rule of empathy). R-beings find their mathematical and scientific knowledge useful for a time, which later they modify as needed. Provisional science is no substitute for the workings of nature, and summary description is no substitute for the work of art.

In our comparison of science and the arts as practiced by r-beings, we have located differences between the two processes. But we have also identified similarities: both processes involve the free imagination of r-beings, and both processes remain forever incomplete.

§8   From Terrestrial Chauvinism to Golden Rule

Will human r-beings in this local region of this universe soon achieve a higher personhood and become advanced, extraterrestrial, transmortal beings? With the previous considerations (of §§1-7) in mind, I will now attempt to articulate ethical-political and other details or implications for r-beings in the historical position humans find themselves embedded in today – as follows:

(A) Perhaps we are in transition from human personhood to advanced personhood.
(B) Perhaps we are in transition from terrestrial personhood to extraterrestrial personhood.
(C) Perhaps we are in transition from mortal personhood to transmortal personhood.

(A) From human beings to advanced beings

Above (in §5) I have distinguished nonpersons from persons in the following way:

• Temporal nonpersonal entities include: Energy (Quanta); Matter (Atoms); and, Life (e.g., Flowers) (Biosystems).

• Temporal personal entities include: Sentience (e.g., Swans) (S-Creatures); and, Reason (R-Beings). Some r-beings are better at reason (reasoning or being reasonable), than others – to wit: Humble Reason (e.g., Human Beings) (H-Beings); and, Advanced Reason (Advanced Beings) (A-Beings).

Although I have spoken above (in §4) of the infinite past and the infinite future, from the specifically human-limited perspective it would seem that almost all of temporal reality is (is to be) located in the future.[11] With reference to our own universe (or our own little corner of it), it seems that non-persons (nonpersonal entities) have dominated our region's known past and that persons (personal entities) may well dominate our region's future. But today's (human-limited) persons are hugely influenced by the past from which they emerged. However, a future person may have the capacity to reinvent oneself, to restructure one's own non-teleological (energy-quanta, matter-atoms, biology-teleonomic) systems and also one's own sentience-hedonic system to conform to the (teleological) results of one's own reasoning or choice, whether moral or immoral, wise or foolish.

Restructuring the energy-system of one's own body might involve advanced subatomic technology as well as insight into reasonable expectations. Restructuring the matter-system, the teleonomic-system, and the hedonic-system of one's own body might involve advanced molecular (nano) technology as well as insight into reasonable expectations. It is of course conceivable that modifying one system might have unknown consequences for the other systems.

I'm not sure we know enough about energy or subatomic technology to yet offer responsible advice about the restructuring of the energy-system of one's own body. However we do have some beginner's insight into the advanced molecular (nano) technology of the future. We may want to begin with modest modifications to our bodies as we gradually learn more. "A little knowledge is a dangerous thing."

Presently I will make a few brief remarks related to the teleonomic-system (biology or life) and the hedonic-system (sentience). The (biology-)teleonomic and (sentience-)hedonic systems of today's human person are structured based on the non-teleological past. This suggests that great changes to these systems are in the long run to be preferred so as to enhance the lives of persons.

Some may believe that a teleonomic system (whether of a rose-flower or of a human) is teleological because it seems to exhibit purposefulness and is goal-oriented. But in fact the teleonomic-system as such is not conscious and is the result of evolutionary adaptation. Although there may be good practical reasons for taking a cautious approach to its modification, from a moral-teleological point of view its improvement is imperative.

Thus in a thought experiment (rather different from our actual world context, or so I believe) we can imagine a world context in which, as a practical matter, there may be good reasons for not extending the healthy lifespan of persons from 50 years to 500 years. In the world in which we actually live, however, my sense is that such so-called reasons are not really very good reasons – we are biased by confusing teleonomy with teleology.

Likewise, many fail to see that our hedonic-system (of pleasure and pain) is also based on the past and should be modified with advice from our system of moral reflection (reasoning). Pleasure and pain, given advanced future technology, could presumably be structured in a wide variety of different ways. (To be sure, a variety of hedonic-systems already exist.) We could structure it so that good behavior is painful and bad behavior is pleasurable. Alternatively, we could structure it so that philosophic reflection and moral behavior are the most pleasurable of pleasures. The point is that "having fun" is neither the only nor highest value, but with future technology we will presumably be able to restructure our system of pleasure and pain to make it more ethical-teleological.

As our own universe (or local region) evolved and became more complex, moral consciousness eventually appeared. Today moral consciousness must learn to unbias or free itself from the teleonomic and hedonic systems of old in order to renovate our blind universe or region. The insect is eating the grass while the mammal is devouring the insect. The mammal, caught in a metal trap, sees the human hunter approaching. Our blind universe or region has cruelly set animal against animal – and humans against mortality.

Here are some of the presumed capacities of advanced beings (or transhumans) as they renovate (well or poorly) our blind universe or region:

> ➤ Use of free-will and great power to pursue wisdom, to learn self-respect, and to respect all persons, past-present-future.

> Insure that no animal kills another animal. This includes both non-human animals (or s-creatures) and human animals (or h-beings).

> Insure that no reflective-person (r-being) must die.

> Insure that no person must experience unwanted serious pain or hardship.

Eventually we may be able to do more than merely retrodict or SIMULATE the past. Eventually we may have the ability to run ancestor history EMULATIONS (via time travel or otherwise). R. Michael Perry (2005) has remarked that it would seem to be immoral to run such ancestor history emulations – real persons would experience real pains and evils. Instead, as Perry advocates, the golden rule would charge us with the duty to revive our ancestors – the scientific resurrection of all dead persons in the omniverse's temporal realm (multiverse of all multiverses).

(B) From terrestrial beings to extraterrestrial beings

The fact that humans presently exist together in a single biosphere global village is a rather absurd position to be in if we seek to prevent doomsday and promote flourishing. [12] If something catastrophic happens to Earth's biosphere, then something catastrophic happens to all Earthlings. It is not wise to put all of humanity's eggs (futures) into one basket (biosphere). "Epitaph: Foolish dinosaurs never escaped Earth."

Advanced Genetic, Robotic, Information, and Nano ("GRIN") technologies are not required for the development of Self-sufficient Extra-terrestrial Green-habitat communities ("SEGs") or independent, self-replicating biospheres in outer space (Seg-communities, 2008). Advanced GRIN technologies certainly will greatly enhance SEG capacities, however.

Self-sufficient Extra-terrestrial Green-habitat communities (SEGs or seg-communities) should not be confused with space stations. Some argue that if we had chosen to do so, we could have started building SEGs using the "merely super" technology of the 20th century. Indeed, the famous 20th century physicist Gerard K. O'Neill designed such SEGs for the purpose of late 20th century construction. Such SEGs would provide a "green-friendly" environment for humans, animals, and plants superior to the problematic habitats we identify with Earth and other planets. In the 20th century the famous physicist Carl Sagan stated: "Our technology is capable of extraordinary new ventures in space, one of which Gerard O'Neill has described to you ... It is practical."

Eventually millions of persons in a single SEG community are possible. The SEGs (seg-communities) would be self-sufficient and could reproduce other SEG habitats in extraterrestrial space at a geometric rate. Accordingly, there is "unlimited free land" in extraterrestrial space – with a higher quality of life than is possible on the surface of a planet.

SEG communities can be built from extraterrestrial resources mined from asteroids or moons. Rotation of the large and spacious greenhouse habitat provides simulated gravity for the people and plants living on the inner surface. Adjustable mirrors provide energy from the sun and simulation of day and night. Sooner or later, the following would be feasible for SEGs:

- "Unlimited energy" from the sun.
  (The sun never sets in space.)

- Control of daily weather and sunlight.

- SEGs would be self-sufficient.

- Expansion of the (self-sufficient) SEGs at a geometric rate.

- "Unlimited free land" via SEGs.
  (Needed raw materials from asteroids are abundant.)

The following metaphorical insights have been widely quoted by SEG experts: "The Earth was our cradle, but we will not live in the cradle forever." "Space habitats [SEGs or seg-communities] are the children of Mother Earth." According to Carl Sagan, our long-term survival is a matter of spaceflight or extinction: "All civilizations become either spacefaring or extinct." According to the "mass extinction" article in *The Columbia Encyclopedia* (6th edition): "The extinctions, however, did not conform to the usual evolutionary rules regarding who survives; the only factor that appears to have improved a family of organisms' chance of survival was widespread geographic colonization." (For us today, we may call this "the extraterrestrial imperative".)

What political philosophy is "fit" for the extraterrestrial imperative? I suggest "PFIT" – P̲eace and F̲reedom, and I̲ntentional T̲ransparent communities, in extraterrestrial space – as follows: What seems to me both practical and fair in this context is to think in terms of a new political philosophy or approach to stable peace in the form of an Extraterrestrial Society of Intentional Communities. There would be two sets of liberties and two sets of responsibilities (for "Extraterrestrial Society" and "Intentional Communities" respectively). Each person is free to found new (intentional) communities. Each Community would determine its own membership requirements. Each Community would have its own culture of liberties and responsibilities; a member would generally be free to leave the community. A mechanism or set of mechanisms would be established to insure that each member is fully and properly informed of their liberty to leave the (intentional) community. (I suppose some communities might still allow their members the possibility of experiencing physical pain – but they would also allow a member to voluntarily leave their community. Too, I suppose banning animal cruelty and serious animal pain would be desirable and feasible.) Note that some ("hermit") communities (SEGs) would consist of only one person.

On old Terra, it was often difficult or impossible to leave one's community – sometimes expulsion effectively meant the individual's death. The context of the Extraterrestrial Society of Intentional Communities is radically different. For example: The individual person would be transmortal, whereas on old Terra it was often the community or society (not the human individual) that was seen as transmortal.

So at the level of the Society (of Communities) we have: (1) Peace: Weapons, weapons-making, and violence (including animal cruelty and serious animal pain) are strongly effectively enforceably banned; and, (2) Freedom: Every individual person is fully aware of and fully informed of their general liberty to leave their community. This too is strongly effectively enforced. The Society and the communities necessarily work closely together to fully insure the liberties and responsibilities associated with both Peace and Freedom. Also note that since there is "unlimited free land," this fact will additionally help prevent some old terra-style conflicts and resolve or manage others (this would include some old-style civil conflicts).

At the level of Communities (in the Society) we have: (1) Intentionality (voluntariness): Within the good-faith transparent enforcement of Society's basic principles of peace and freedom, each Community has wide latitude for experimentation. Although there is a general liberty of members to leave the (intentional) Community, this does not necessarily relieve such persons from certain good-faith responsibilities to the Community; and, (2) Transparency: Each Community must strongly, effectively, and transparently help enforce the Society's basic principles of peace and freedom.

I believe the political theory or moral-political approach I have invented above is unique and original. It differs from the "Law of Peoples" conception of John Rawls (1999) in that it primarily chooses a "Law of Persons" model instead. Yet it takes seriously the distinction Rawls makes between a "political conception" and "comprehensive doctrines." In my "PFIT" or "Society of Communities" theory, Society corresponds to a political conception or model, and Communities (SEGs) represent comprehensive doctrines or worldviews.

Like Charles R. Beitz (1999), my theory takes seriously a cosmopolitan-political "Law of Persons" (as distinguished from a social-political "Law of Peoples") approach. It differs from Beitz in methodology and in the questions asked. Beitz finds the question of distributive justice both highly important and practically difficult with respect to present Terrestrials. This is a question I do not raise since in my extraterrestrial world of the future it seems not an issue or one rather resolvable in that easier context of expanded liberty – there requiring perhaps at most only a bit of good-will and ingenuity.

"Is stable peace possible if each person or each people is passionately convinced their worldview is basically good and correct – and different worldviews are evil or bad or incorrect?" If you can sincerely and in good faith agree to my political approach above, the answer to this question appears to be YES, such stable peace is possible. If you can at most only agree to my approach as a temporary compromise, then the answer may be NO.

"If we could enforceably prevent each and every person from killing any person over a conflict (say, a conflict of worldviews) would we do so? If so, how would we resolve our conflicts?" If you can sincerely and in good faith (instead of merely as a temporary compromise) agree to my approach above, then stable peace in extraterrestrial space seems both possible and desirable. This approach, so I believe, realistically outlines a structure of stable peace for World Society and local Communities (SEGs or seg-communities) in extraterrestrial space – pointing toward conflict management in the new framework and encouraging subsequent projects to invent needed specifics.

The first (temporary) Extraterrestrial Space Treaty seems doable today. A permanent Extraterrestrial Space Treaty seems doable soon. A Universal Space Treaty that includes both Extraterrestrial Space and Terrestrial Space may take more time but appears to be a goal worth striving for – indeed, the striving itself may well improve matters. In the meantime, the previous treaties and upward strivings should make these "final strivings" toward a Good Society more nearly achievable for all.

(C) From mortal beings to transmortal beings with a common task?

According to the omniverse model presented above, any purely physical-scientific account of reality must be deficient. I believe my general-ontological framework should prove fruitful when discussing or resolving philosophic controversies – and helpful to scientists and lay folks as well. The topic we now turn to is the question of personal immortality. On this issue, the "golden goodwill" of A. N. Whitehead (1929) (1941), Albert Camus (1942) (1951), and the omniverse model triumph over the "midnight madness" of Martin Heidegger (Shaviro, 2009) [13] and numerous others. [14]

Jacques Choron (1973: p. 638) notes that: "The main difficulty with personal immortality ... is that once the naive position which took deathlessness and survival after death for granted was shattered, immortality had to be proved. All serious discussion of immortality became a search for arguments in its favor." "In order to be a satisfactory solution to the problems arising in connection with the fact of death, immortality must be first a 'personal' immortality, and secondly it must be a 'pleasant' one."

How shall we deal with the apparent conflict between immortality and entropy? According to the omni paradigm, is entropy a fake? Note that the "dismal" theory of thermodynamics in the form of its second law (the so-called "entropy" law) applies to closed/isolated systems. But given the context of the omniverse model (see §7 above), we can now say: Gödel showed us (if we did not know it already) that all-reality (the omniverse) is not a closed/isolated system. "The entropy concept," according to Kenneth Boulding (1981: p. 10), "is an unfortunate one, something like phlogiston (which turned out to be negative oxygen), in the sense that entropy is negative potential. We can generalize the second law in the form of a law of diminishing potential rather than of increasing entropy, stated in the form: If anything happens, it is because there was a potential for it happening, and after it has happened that potential has been used up. This form of stating the law opens up the possibility that potential might be re-created."

Again I emphasize that the second law does not really say that (all-reality's) potential is finite. Instead, let me suggest that the second law may be related to the arrow of time or to the fact that "Once I do X instead of Y, X will always be the case."

Work beginning in the 20th century has laid the foundation for eventual realization of transmortality and more, the onto-resurrection imperative or common task of resurrecting all past persons no longer alive. Developments have already taken us to the threshold of what has been called "practical time travel" – or what, loosely speaking, we may call "time travel": See §4 above. Once time travel becomes feasible in the 21st century, then we can proceed to more fully implement our common task of resurrecting all dead persons (rather than some dead persons via cardiopulmonary resuscitation). The first steps occurred in the 20th century on several fronts, including steps in the direction of suspended-animation, superfast-rocketry, and seg-communities. [15]

Experts tell us that the results of the population explosion (i.e. the size of the human population) will level off sometime in the 21st century (perhaps mid-century). Experts also tell us that current and ongoing industrial-technological activities are dangerously polluting our planet and causing global warming; global warming, in turn, can very easily lead to unprecedented injustices and upheavals in a terror-filled global-village of weapons of mass death and destruction. Presumably we should take global action against global dangers along the lines suggested by Al Gore, Jared Diamond, and other experts; see the Gore-related website about the practical generation of carbon-free electricity: www.RepowerAmerica.org; also see the Diamond-related website about "the world as a polder": www.mindfully.org/ Heritage/2003/Civilization-Collapse-EndJun03.htm. But certainly too we can and should engage in additional terrestrial and extraterrestrial activities to prevent doomsday and improve the human condition. If we are not balanced and careful in our actions, myopia can provide us with badly-needed near-term clarity while preventing us from the broader vision required for survival, thrival, and the common task.

Perfection of future-directed time travel in the form of suspended-animation (biostasis) seems feasible in the 21st century. I believe it even seems feasible to eventually offer it freely to all who want it. Jared Diamond (2005: p. 494) has pointed out that: "If most of the world's 6 billion people today were in cryogenic storage and neither eating, breathing, nor metabolizing, that large population would cause no environmental problems." [16] Too, this might allow them to travel to an improved world in which they would be transmortal. Since aging and all other diseases would have been conquered, they might not have to use time travel again unless they had an accident requiring future medical technology.

The onto-resurrection imperative demands more than immortality for those currently alive. In extraterrestrial space we can experiment (perhaps, for example, via past-directed time travel-viewing) with immortality for all persons no longer alive. Seg-communities (Self-sufficient Extra-terrestrial Green-habitats, or O'Neill communities) can assist us with our ordinary and terrestrial problems as well as assist us in completion of the onto-resurrection project. Indeed, in Al Gore's account of the global warming of our water planet, his parable of the frog is a central metaphor. Because the frog in the pot of water experiences only a gradual warming, the frog does not jump out. I add: Jumping off the water planet is now historically imperative; it seems unwise to put all of our eggs (futures) into one basket (biosphere).

With respect to our common task (the onto-resurrection imperative), I quote Jacques Choron (1973) once again: "Only pleasant and personal immortality provides what still appears to many as the only effective defense against ... death. But it is able to accomplish much more. It appeases the sorrow following the death of a loved one by opening up the possibility of a joyful reunion ... It satisfies the sense of justice outraged by the premature deaths of people of great promise and talent, because only this kind of immortality offers the hope of fulfillment in another life. Finally, it offers an answer to the question of the ultimate meaning of life, particularly when death prompts the agonizing query [of Tolstoy], 'What is the purpose of this strife and struggle if, in the end, I shall disappear like a soap bubble?'" (p. 638)

§9   Closing Remarks

An outline of reality, herein called the "omni" (omni-universe or omniverse) model, has been presented and justified. The paper discussed the nature and obligations of temporal personal entities with the ability to reason and be reasonable ("r-beings") in an omniverse environment. R-beings with the limited reason of humans have an obligation to become advanced r-beings, and advanced r-beings have an obligation to advance further and further. As r-beings advance, they outgrow the chauvinism of my-species and my-planet. With perpetually advancing knowledge gained from scientific method and golden rule, r-beings are able to improve world and self.

With this in mind, the paper articulated ethical-political and other details or implications for r-beings in the historical position humans find themselves today. A political philosophy "fit" for the extraterrestrial imperative was suggested: "PFIT" – i.e., Peace and Freedom, and Intentional Transparent communities, in extraterrestrial space. This new political philosophy was explained and defended. PFIT is believed to be a feasible approach to achieving stable peace in the form of an Extraterrestrial Society of Intentional Communities (seg-communities).

The paper showed that all-reality, or the infinity of infinities I have called the omniverse, is not altogether reducible to any strictly physical-scientific paradigm. A more believable (general-ontological) paradigm was presented. Within this framework, the issue of personal immortality was considered. It was concluded that the immortality project, as a physical-scientific common-task to resurrect all dead persons, is ethically imperative. The imperative includes as first steps the development of successful antiaging-methods, longterm suspended-animation, Einsteinian superfast-rocketry, and PFIT seg-communities.

As r-beings learn more and more about the infinite game, presumably they will eventually learn how to engender universes with laws tailored to their specifications – and intervene contrary to those laws. Presumably they could also use those laws and interventions to produce strongly convincing virtual realities or appearances in apparent contradiction to "natural laws" – and more, even "contradicting" the laws of mathematics and logic. This is perhaps just the sort of thing Grand Magicians or Advanced Beings or Magisters Ludi would enjoy doing. Shall we continue to continue to continue … playing the infinite game?

One difficulty in playing the infinite game is our huge (infinite) ignorance of the infinite omniverse both temporal and non-temporal. The paper's analysis of the situation of r-beings (both human and advanced) in the omniverse environment finds some analogy to "the veil of ignorance" that is (hypothetically) posited by John Rawls (1971) (1999) for the purpose of identifying proper or just political arrangements. It apparently turns out that in the omniverse environment, a kind of veil of ignorance is real, not just hypothetical, for both human beings and advanced beings. Thus it seems that there is now a real motivating force for "us" (whether human-beings, advanced-beings, advanced advanced-beings, et cetera ad infinitum) to behave toward each other in a "golden" way. It also seems that (Rawlsian or political) "justice" is only one "value" among an infinity of "coordinated values" in the paragonal realm of necessary reality. This coordinated "infinity of paragonals" we may call "the Paragon" or "the Necessary" or "the Required" or "the Good".

Acknowledgements

I would like to thank the philosophy department of National Central University (Taiwan) for their "Symposium on Environmental Ethics" interaction, and the philosophy department of National Chung Cheng University (Taiwan) for their "Visiting Scholar" assistance. I would also like to thank William Grey and J. R. Lucas for their critical and helpful comments. Adapted from (Tandy, 2009b).

Bibliography

Barzun, Jacques and Graff, Henry F. (1985). *The Modern Researcher: Fourth Edition*. Harcourt Brace Jovanovich: San Diego.

Bergson, Henri (1932). *The Two Sources Of Morality And Religion*. Translated by R. Ashley Audra and Cloudesley Brereton with the assistance of W. Horsfall Carter. University Of Notre Dame Press: Notre Dame. (1932; 1935; this translated edition, 1977).

Best, Benjamin P. (2008). "Scientific Justification of Cryonics Practice" *Rejuvenation Research* vol. 11, no. 2 [April 2008]: Pages 493-503. Also available at <www.cryonics.org/reports/Scientific_Justification.pdf>.

Beitz, Charles R. (1999). *Political Theory and International Relations: With a New Afterword by the Author*. Princeton University Press: Princeton, NJ.

Bostrom, Nick (2003). "Are You Living In A Computer Simulation?" *Philosophical Quarterly* 53(211) [2003]: Pages 243-255. Also see a Nick Bostrom website: <http://www.simulation-argument.com>.

Bostrom, Nick and Roache, Rebecca (2007). "Three Big Problems" Pages 147-164 In: Tandy, Charles [Editor] (2007). *Death And Anti-Death, Volume 5: Thirty Years After Loren Eiseley (1907-1977)*, Ria University Press: Palo Alto, California (USA). (ISBN 9781934297025).

Boulding, Kenneth E. (1956). *The Image*. University of Michigan Press: Ann Arbor. (This edition, 1961, Ann Arbor Paperback).

Boulding, Kenneth E. (1981). *Ecodynamics: A New Theory of Societal Evolution*. Sage Publications: Beverly Hills. (First edition, 1978; this edition, 1981).

Bronowski, Jacob (1966). "The Logic of Mind" *American Scientist*, 54 (1), March 1966, Pages 1-14. This is approximately reprinted as the "supplement" chapter in Bronowski (1971).

Bronowski, Jacob (1971). *The Identity of Man*. Natural History Press: New York. This is the revised and expanded 1971 (not 1965) edition. The new "supplement" chapter serves as an approximate reprint of Bronowski (1966).

Burtt, E. A. (1965). *In Search Of Philosophic Understanding*. New American Library: New York. (1967 Edition).

Camus, Albert (1942). *The Myth of Sisyphus*. Vintage Books: New York. (Originally published in French in 1942). This translated edition, 1991.

Camus, Albert (1951). *The Rebel: An Essay on Man in Revolt*. Vintage Books: New York. (Originally published in French in 1951). This translated edition, 1991.

Catterson, Troy T. (2003). "Letting The Dead Bury Their Own Dead: A Reply To Palle Yourgrau" Pages 413-426 In: Tandy, C. [Editor] (2003). *Death And Anti-Death, Volume 1*. Ria University Press: Palo Alto, CA.

Chaitin (1982). Gregory J. Chaitin. "Gödel's Theorem and Information", *International Journal of Theoretical Physics*, 21, [1982], Pages 941-954.

Choron, Jacques (1973). "Death and Immortality" in Volume 1 (Pages 634-646) of *The Dictionary of the History of Ideas* edited by Philip P. Wiener. (1973=vols.1-4; 1974=index vol.). Charles Scribner's Sons: New York. Available at <http://etext.virginia.edu/DicHist/ dict.html>.

De Grey, Aubrey D.N.J. (2007). "Is It Safe for a Biologist to Support Cryonics Publicly?" Pages 235-258 In: Tandy, Charles [Editor] (2007). *Death And Anti-Death, Volume 5: Thirty Years*

After Loren Eiseley (1907-1977), Ria University Press: Palo Alto, California (USA). (ISBN 9781934297025).

Descartes, René (1637). ***Discourse on the Method***. (Originally published anonymously in French, 1637). (Various translations available).

Diamond, Jared (2005). ***Collapse: How Societies Choose to Fail or Succeed***. Viking: New York.

Eiseley, Loren (1973). ***The Man Who Saw Through Time***. Charles Scribner's Sons: New York. (1961, 1962, 1964). This edition, 1973. [Francis Bacon is "The Man"].

Ettinger, R. C. W. (2002). "Youniverse" Pages 237-272 In: Tandy, C. And Stroud, S. R. [Editors] (2002). ***The Philosophy Of Robert Ettinger***. Ria University Press: Palo Alto, CA.

Ettinger, R. C. W. (2004). "To Be Or Not To Be: The Zombie In The Computer" Pages 311-338 In: Tandy, C. [Editor] (2004). ***Death And Anti-Death, Volume 2***. Ria University Press: Palo Alto, CA.

Ettinger, Robert C. W. (2005). ***The Prospect of Immortality***. Ria University Press: Palo Alto, CA. [Privately published 1962, Doubleday edition 1964, subsequent editions in several languages; this edition, 2005].

Fedorov, Nikolai Fedorovich (2008). [Two websites about him:] <http://www.iep.utm.edu/f/fedorov.htm>; and, <http://www.quantium.plus.com/venturist/fyodorov.htm>.

Feigl, Herbert (1958). "The 'Mental' and the 'Physical'" in ***Minnesota Studies in the Philosophy of Science: Volume II: Concepts, Theories, and the Mind-Body Problem*** edited by Herbert Feigl, Michael Scriven, and Grover Maxwell. University of Minnesota Press: Minneapolis. (Pages 370-497). (See especially section "V.c." on pages 431-438).

Gödel, Kurt (1931). "Über Formal Unentscheidbare Sätze der *Principia Mathematica* und verwandter Systeme [Part] I" *Monatschefte für Mathematik und Physik*, Volume XXXVIII, [1931], Pages 173-198. (Reprinted with English translation in *Kurt Gödel: Collected Works*, Volume 1, Oxford University Press: New York, 1986, Pages 144-195).

Gore (2006). Al Gore. *An Inconvenient Truth: The Planetary Emergency of Global Warming and What We Can Do About It*. Rodale Books: Emmaus, Pennsylvania. [This is the first book in history produced to offset 100% of the CO_2 emissions generated from production activities with renewable energy; this publication is carbon-neutral.]

Hartshorne, Charles (1941). *Man's Vision of God and the Logic of Theism*. Willett, Clark and Co.: New York. This is an expanded treatment of Hartshorne (1951) and more. [Perhaps Hartshorne (1951) was first published in 1941?].

Hartshorne, Charles (1951). "Whitehead's Idea of God" Pages 515-559 In: Schilpp, Paul Arthur (editor) (1951). *The Philosophy of Alfred North Whitehead: Second Edition*. Open Court: La Salle, Illinois. [Perhaps this was first published in 1941 in the First Edition?].

Hartshorne, Charles (1962). *The Logic of Perfection*. Open Court: La Salle, Illinois.

Jackson, Frank (1982). "Epiphenomenal Qualia", *Philosophical Quarterly*, XXXII (32), April 1982, Pages 127-136.

Jackson, Frank (1986). "What Mary Didn't Know", *Journal of Philosophy*, LXXXIII (83), May 1986, Pages 291-295.

Kierkegaard, Soren (1847). *Works Of Love*. Translated by H. Hong and E. Hong. Harper & Row: New York. (1962, 1964). [In his journal, Kierkegaard wrote: "Subjectivity is truth."].

Kuhn, Thomas (1962). *The Structure of Scientific Revolutions*. University of Chicago Press: Chicago. (Second edition enlarged, 1970).

Lepore, E. and Van Gulick, R. [Editors] (1991). *John Searle And His Critics*. Basil Blackwell: Oxford.

Leslie (2007). John Leslie. *Immortality Defended*. Blackwell Publishing: Oxford.

Li, Jack (2002). *Can Death Be a Harm to the Person Who Dies?* Kluwer Academic Publishers: Dordrecht/Boston/London. [See especially chapter four for a defense of objective interests (objective values).]

Lucas, J. R. (2008). "[Section:] Gödelian Arguments" at his <http://users.ox.ac.uk/~jrlucas/reasreal/reaschp6.pdf>.
This section is in chapter two of Lucas (2009).

Lucas, J. R. (2009). *Reason and Reality*. Ria University Press: Palo Alto, CA.

Nagel, Ernest and Newman, James R. (1958). *Gödel's Proof*. Routledge: London. (First edition, 1958; this edition, 2002).

Nagel, Thomas (1974). "What Is It Like to Be a Bat?" *Philosophical Review*, LXXXIII (83), 4 (October 1974), Pages 435-450.

O'Neill, Gerard K. (2000). *The High Frontier: Human Colonies in Space*. Apogee Books: Burlington, Ontario, Canada. (3rd edition). Also see: <http://www.space-frontier.org/HighFrontier/testimonial.html>.

Orwell, George (1949). *1984*. New American Library: New York. (First edition, 1949; this edition, 1961).

Penrose, Roger (1989). *The Emperor's New Mind*. Oxford University Press: New York.

Penrose, Roger (1990). "Précis", *Journal of Behavioral and Brain Sciences*, 13 (4), [1990], Pages 643-654.

Penrose, Roger (1994). *Shadows of the Mind*. Oxford University Press: New York.

Penrose, Roger (2005). *The Road to Reality: A Complete Guide to the Laws of the Universe*. Alfred A. Knopf: New York. (First edition, 2004; this edition, 2005).

Perry, R. Michael (2000). *Forever For All: Moral Philosophy, Cryonics, And The Scientific Prospects For Immortality*. Universal Publishers: Parkland, FL.

Perry, R. Michael (2005). Personal Communication From R. Michael Perry To Charles Tandy (21 August 2005).

Rawls, John (1971). *A Theory Of Justice*. The Belknap Press Of Harvard University Press: Cambridge, MA. Revised Edition, 1999. (Original Edition, 1971).

Rawls, John (1999). *The Law of Peoples: with "The Idea of Public Reason Revisited"*. Harvard University Press: Cambridge, MA. This edition, 2001.

Rickert, Heinrich (1902). *The Limits of Concept Formation in Natural Science*. Translated and abridged by Guy Oakes. Cambridge University Press: Cambridge, 1986. [First published in German in 1902.] [Rickert defends valid values (objective values).]

Scientists' Open Letter on Cryonics (2006). [61 signatories: "Cryonics is a legitimate science-based endeavor … " etc., plus bibliography]. Pages 247-258 In: Tandy, Charles [Editor] (2007). *Death And Anti-Death, Volume 5: Thirty Years After Loren Eiseley (1907-1977)*, Ria University Press: Palo Alto, California

(USA). (ISBN 9781934297025). Also available at <www.imminst.org/cryonics_letter/index.html>.

Searle, J. (1980). "Minds, Brains And Programs." *The Behavioural And Brain Sciences* 3 [1980]: Pages 417-57.

Searle, J. (1984). *Minds, Brains And Science*. Harvard University Press: Cambridge, MA.

Sen, Amartya (1999). *Development as Freedom*. Anchor Books: New York. This edition, 2000.

Seg-communities (2008). <www.ria.edu/seg-communities>. [Or see these six websites about seg-communities (Self-sufficient Extra-terrestrial Green-habitats, or O'Neill communities):]
(1) <http://en.wikipedia.org/wiki/Space_colonization>;
(2) <http://www.nas.nasa.gov/About/Education/SpaceSettlement>;
(3) <http://www.nss.org/settlement/space/index.html>;
(4) <http://www.segits.com>;
(5) <http://www.spaext.com>; and,
(6) <http://www.ssi.org>.

Shaviro, Steven (2009). *Without Criteria: Kant, Whitehead, Deleuze, and Aesthetics*. MIT Press; Cambridge, MA.

Strawson, P. F. (1958). "Persons" in *Minnesota Studies in the Philosophy of Science: Volume II: Concepts, Theories, and the Mind-Body Problem* edited by Herbert Feigl, Michael Scriven, and Grover Maxwell. University of Minnesota Press: Minneapolis. (Pages 330-353).

Strawson, P. F. (1959). *Individuals: An Essay In Descriptive Metaphysics*. Routledge: London. (1959; 1964; this edition,1990).

Strawson, P. F. (1985). *Scepticism and Naturalism: Some Varieties*. Columbia University Press: New York.

Tandy, Charles (2002). "Toward A New Theory Of Personhood" Pages 157-188 In: Tandy, C. [Editor] (2002). *The Philosophy Of Robert Ettinger*. Ria University Press: Palo Alto, CA.

Tandy, Charles (2003). "N. F. Fedorov And The Common Task: A 21st Century Reexamination" Pages 29-46 In: Tandy, C. [Editor] (2003). *Death And Anti-Death, Volume 1*. Ria University Press: Palo Alto, CA.

Tandy, Charles (2006). "A Time Travel Schema And Eight Types Of Time Travel," In Tandy, Charles [Editor] (2006). *Death And Anti-Death, Volume 4: Twenty Years After De Beauvoir, Thirty Years After Heidegger*, Ria University Press: Palo Alto, California (USA). (ISBN 9780974347288). (Pages 369-388).

Tandy, Charles (2007a). "Types Of Time Machines And Practical Time Travel" *Journal Of Futures Studies* vol. 11, no. 3. [February 2007]: Pages 79-90. Available at <http://www.jfs.tku.edu.tw/11-3/A05.pdf>.

Tandy, Charles (2007b). "Teleological Causes And The Possibilities Of Personhood," In Tandy, Charles [Editor] (2007). *Death And Anti-Death, Volume 5: Thirty Years After Loren Eiseley (1907-1977)*, Ria University Press: Palo Alto, California (USA). (ISBN 9781934297025). (Pages 399-416).

Tandy, Charles (2007c). "Terrestrial Peoples, Extraterrestrial Persons," In Tandy, Charles [Editor] (2007). *Death And Anti-Death, Volume 5: Thirty Years After Loren Eiseley (1907-1977)*, Ria University Press: Palo Alto, California (USA). (ISBN 9781934297025). (Pages 417-432).

Tandy, Charles (2008). "What Mary Knows: Actual Mentality, Possible Paradigms, Imperative Tasks," In Tandy, Charles [Editor] (2008). *Death And Anti-Death, Volume 6: Thirty Years After Kurt Gödel (1906-1978)*, Ria University Press: Palo Alto, California (USA). (ISBN 978-1-934297-03-2). (Pages 265-284).

Tandy, Charles (2009a)."Entropy And Immortality," ***Journal Of Futures Studies***, Volume 14, Number 1 (August 2009). (ISSN 10276084). (Pages 39-50). Available at <http://www.jfs.tku.edu.tw/14-1/A03.pdf>.

Tandy, Charles (2009b). "Personal, Temporal, And Paragonal Aspects Of The Omniverse," In Tandy, Charles [Editor] (2009). ***Death And Anti-Death, Volume 7: Nine Hundred Years After St. Anselm (1033-1109)***, Ria University Press: Palo Alto, California (USA). (Hardback/ Hardcover ISBN 978-1-934297-05-6). (Paperback/ Softcover ISBN 978-1-934297-07-0). (Chapter 14).

Time-travel (2008). <www.ria.edu/time-travel>. [Or see these websites about (1) time-travel; (2) suspended-animation; and, (3) superfast-rocketry:]
(1) <http://www.jfs.tku.edu.tw/11-3/A05.pdf>;
(2) <http://en.wikipedia.org/wiki/Greg_Fahy>; and,
(3) <http://en.wikipedia.org/wiki/Twin_paradox#Resolution_of_the_paradox_in_general_relativity>.

Transhumanism (2008). [Two transhumanist websites:]
(1) <http://www.aleph.se/Trans>; and,
(2) <http://www.transhumanism.org>.

Waddington, C. H. (1967). ***The Ethical Animal***. University Of Chicago Press: Chicago.

Whitehead, Alfred North (1929). ***The Function of Reason***. Beacon Press: Boston, MA. (This printing, 1959).

Whitehead, Alfred North (1941). "Immortality" Pages 682-700 In: Schilpp, Paul Arthur (editor) (1951). ***The Philosophy of Alfred North Whitehead: Second Edition***. Open Court: La Salle, Illinois.

Endnotes

1. §5 is based on Tandy (2007b), pages 407-409.

2. I have here paraphrased Whitehead (1929): See especially chapter 1.

3. The following five paragraphs are either based on, or inspired by, chapter 1 of Whitehead (1929).

4. The following four paragraphs are either based on, or inspired by, chapter 2 of Whitehead (1929).

5. This paragraph and the following two paragraphs are either based on, or inspired by, chapter 3 of Whitehead (1929).

6. The following four paragraphs are either based on, or inspired by, chapter 1 of Bronowski (1971).

7. This paragraph and the following six paragraphs are either based on, or inspired by, chapter 2 of Bronowski (1971).

8. The following six paragraphs are either based on, or inspired by, chapter 3 of Bronowski (1971).

9. The following five paragraphs are either based on, or inspired by, chapter 4 of Bronowski (1971).

10. This paragraph and the following six paragraphs are either based on, or inspired by, the "supplement" chapter (the final chapter, following chapter 4) of Bronowski (1971). The "supplement" chapter serves as a reprint of Bronowski (1966).

11. This paragraph and the following eight paragraphs are based on Tandy (2007b).

12. This paragraph and the following 14 paragraphs are based on Tandy (2007c). Also see: Seg-communities (2008).

13. Shaviro (2009) focuses on the question of "what if" 20th century philosophy had taken Heidegger less seriously and Whitehead more seriously.

14. This paragraph and the following eight paragraphs are based on Tandy (2008).

15. See: Time-travel (2008); and, Seg-communities (2008).

16. This may be an exaggeration in that the production of liquid air/nitrogen requires energy; even so, Diamond would appear to be mostly correct here. But it is also conceivable that all or almost all power plants and related technologies will become carbon-neutral or even carbon-extracting. For example, see one of "Al Gore's websites" related to the practical generation of carbon-free electricity: <www.RepowerAmerica.org>. (Some environmentalists say that the additional step or capacity of carbon-extraction is required – or is at least desirable to make our lives easier. Whether practical carbon-extraction techniques would or would not require advanced molecular nanotechnology is not immediately obvious to me. Whether carbon-extraction, carbon-offsets, weather-modification, or terra-forming might be used as a doomsday weapon or weapon of mass death and destruction is yet another matter.)

CHAPTER TWELVE

The Non-human Natural World, Indigenous Peoples, and Late-modern Capitalism

康柏

Mac Kang Bai (Campbell)

Equity Value Governance

> ... your sheep that were wont to be so meek and tame, and so small eaters, now, as I heard say, be become so great devourers and so wild, that they eat up, and swallow down the very men themselves. [1]

Imagine you are a member of an indigenous group dependent on the natural environment. Your meaning system, your language, your laws, are intimately connected with the natural world. The strange faces you see are a survey party, followed by a thing called a fence, something you have never seen or imagined – a straight-line-thing. Imagine that new people kill or chase off any of your people who cross the straight-line-thing. More and more straight-line-things appear, and more and more of your people are killed.

You learn what the straight-line-things mean. You are told that the killings have happened because you are under a new group of strange obligations, set up by a new entitlement made in the name of someone far away. You are told that a man who has never been here, has been imputed with an entitlement by another man who has never been here, using a scribbling called a title deed.

It is this title deed, it is said, that is the reason the strange ones use overwhelming force to ensure that you subordinate your interests, comply with your new obligations, acquiesce in the fouling by strange beasts of your sacred waterholes, acquiesce in the destruction of your way of life, acquiesce in the violation of your country. Far-away scribbles authorize the killing, enslavement, and marginalization of your people.

Those scribbles are the cyber-engines of capitalism. Those scribbles, those title-deeds, set up enforced relations of entitlement and obligation arising out of an artificial relation, an imputed official relation between a person or corporation, and an alienated part of the non-human natural world. Those scribbles are the cyber-engines of capitalism, and they have a history.

I propose now to use an inquiry into the regulatory treatment of indigenous peoples in British colonies, to further consider the limitations of political philosophy in the work of governance. I intend to make a history; a history of the displacement of one form of governance by another. The history of human regulatory arrangements is the history of emergent adaptations to changed circumstances. All human regulatory arrangements are hybrids, developed from previous arrangements by invention, accident, and selection under pressure of demand. The purpose of showing from history how new regulatory traditions emerged is in part to demonstrate just how small has been the influence of political philosophy, compared with the role of accident and unintended consequence under pressure of circumstance, in particular the problems of social order and scarcity.

Those with even a superficial familiarity with the historical enterprises of the schools of insight into the nature of capitalism, will immediately recognize the similarity of what I am attempting; to their emphases on how commodification came to prevail in traditions of governance.

Where others might use the word capitalism, I tend to prefer the term equity value, because I think that equity value comes closer to the heart of the matter, and has a precise meaning. Equity value is the amount a lender will lend in respect of an entitlement created and sustained by governance. The change I want to track is the displacement of practices of governance suited to prior models of engagement with the non-human natural world. The dominant model of relating the non-human and the human worlds in the colonial tradition is a model of equity entitlement, that is to say, a model of governance dominated by equity value. In tracing this contingent development I will trace the emergence of the domination of the meanings of governance by equity value in land.

The following extended section genealogizes late-modern equity value governance. My analysis owes debts to the thinking of Hernando de Soto, Andro Linklater, Fred D'Agostino, and Paul Ivory.[2] Before going on I wish to make some claims about the importance of what I am about to do. Cultures differ in their answers to the two meta-questions asked by philosophers: *What is the nature of reality?* and *How then shall we live?* In order to deal with this I am committed to making some (hopefully uncontroversial) claims about the survival and welfare of past societies. My tools of analysis of past societies are worth noting in advance. I understand governance as the creation and sustaining of entitlement. This genealogy analyses the emergence of new traditions, of new commensurations. Commensuration is a tradition of making operational judgments in situations where two or more incommensurate scales of measurement must be taken into consideration. New traditions of judgment develop by various means; by invention, by incremental steps, and sometimes by accident of circumstance, but once developed they are passed on. So my analysis of the emergence of traditions of commensuration stumbles along, as histories do, knocking up against accident, adaptation and invention. I am also committed to making specific recommendations. Here I part company with those analysts who make none, and also differ, I hope, from those whose recommendations are so broad as to be useless.

I will assert that equity value in entitlement to land is a complex social construct, a product of numerous inventions and adaptations, and not part of the fabric of the universe. I will demonstrate that British colonial governance was the *sine qua non* of emergent equity value creation in the colonies, *via* the surveying and official alienation of the non-human natural world standardized as land. This feature came so to dominate the *telos* of British colonial regulatory procedures that they may usefully be called equity value governance. I go on to trace in broad outline the historical results for colonized indigenous peoples, of equity value governance.

This essay draws attention to the interests of the state as they show up in the design of procedures, with particular focus on procedures whereby indigenous citizens bring claims to control land. Official arrangements in polities which were once British colonies have undergone many changes since first settlement, but between institutions then and now there are similarities important to this essay. I therefore propose to trace the histories of the institutions and technologies of regulation used by the British in their colonies. The purpose of this section is to frame an earlier historical picture of the meanings of British governance as they changed over time, with special reference to its results for indigenous peoples.

The Mabo decision of the High Court of Australia decided that indigenous law was not extinguished by the arrival of British law.[3] At the level of consciousness, not to mention policy, this decision came two centuries too late both for the colonizers and the colonized. Whilst I lack both the resources and the social standing in indigenous communities that would be necessary to attempt to trace the elements of indigenous laws in the colony of New South Wales two centuries ago, there is no difficulty in working up a Western historical picture of the meanings of British governance in the colonies. As an historian I offer here a picture in part influenced by research done by others, but an original picture for all that. By way of example of prior historical investigation, there has arisen a body of work about terms of justification such as *terra nullius and res nullius*. Much of that literature has been by way of careful examination of texts, producing evidence of the thinking of early colonizers. This essay, however, seeks the British colonial *telos* through another approach. That approach is through the examination of institutions. I propose to show that the earlier development of the enabling technologies of symbiosis between British governance and equity value, was in part accidental, but developed under pressure of interest and demand within an institutional tradition, an institutional *telos*. I propose to demonstrate that the symbiosis we might call late-modern equity value governance, developed *via* a number of inventions and adaptations against a particular institutional backdrop. What follows therefore, is an account of those inventions and adaptations, and of their effects.

Regulatory arrangements tend to track the effects of technological adaptations, as I propose to show in the following picture of adaptations in the British regulatory tradition. I propose to use a broad brush in describing the effects of adaptations as they shaped colonial possibilities. Bearing in mind that technologies tend to be interconnective in their effects, I propose to survey the effects of five technologies. They are the technologies of incorporation, of audit, of violence, of surveying and equity value in tandem, and technologies of finance. Because of its highly selective nature, this survey will, in places, be afflicted with the problems of compression in writing history; there are always significant counter-trends and exceptions that suffer omission. I freely admit that the following compressed history of British instruments of entitlement suffers from problems that every historian would like to see remedied at length. Despite these shortcomings, my purpose is not dishonestly selective, but rather an attempt to impress my reader with a proper humility concerning inherited arrangements, with an awareness that arrangements are not part of the fabric of the universe, and might well have been otherwise.

The first technology which we will consider is a good illustration of the historical ubiquity of symbiosis between wealth and governance. This symbiosis is central to my descriptive analysis of the historic relegation of indigenous value systems in the colonial period. In most civilizations throughout history, regulators have been in bed with powerful interests, and so officially promulgated and enforced entitlements have tended to reflect the interests of the powerful at the expense of the interests of the weak. "The powerful do what they will, the weak do as they must."[4] This is historically obvious in the balance of entitlements, and in the designs of regulatory systems which produce entitlements. When we observe historical change in the designs of regulatory arrangements, it is usually safe to say that the change tends to favor the strong over the weak, up to but usually not beyond the point at which social control and prosperity are put at risk. This essay proposes that such a state of affairs, whilst historically normal, is not necessarily in the long-term interests of the state.

This first technology is the convention that a group could be dealt with under the legal fiction that it might properly be treated as though it were one person. This device was familiar to English regulators from medieval times, when collective municipal identity was acknowledged by the granting of charters of incorporation.[5] The usefulness of the device to enable entitlement beyond the scope of individuals, developed to become a major component in the history of commercial speculation.

Now we will consider technologies of audit. Institutions normally place their agents under conditions of audit, so that their official behaviors under audit must reflect the *telos* and the values of the institution and its technologies, regardless of their private opinions. Regulatory arrangements under conditions of audit produce social possibilities and impossibilities which may be described as expressions of an institutional set of understandings of the nature of reality, and of what is best for civilization. These are expressed in an institutional *telos*, an institutional axiology, an institutional ontology, an institutional epistemology, which together emerge as institutional practices do, often by unplanned developments. I emphasize that these institutional presuppositions about the nature of reality have a life and a logic of their own, independent of the individual thinking of the regulators themselves. Regulators, where their private opinions, values, attitudes, beliefs, and purposes differ from official policy, must act as agents of the institutional *telos*. Their hands are tied. This is an important feature of institutional regulatory practice.

Since their adoption of the Chinese inventions of the stern-post rudder and the magnetic compass, European powers progressively improved techniques of navigation under the pressure of the great costs in shipping losses.[6] The colonial incentives of trade and markets were so compelling to private investment that by the close of the eighteenth century new technologies brought navigation and mapping to a high state of reliability.[7] These improvements had been wholeheartedly supported by the British Admiralty and the Royal Society. The effect was that mail throughout the British colonies was sufficiently reliable and timely that audit of colonial functionaries became a given. The effect was to cause even those colonial regulators with wide discretionary powers to execute their duties according to their office, rather than acting on their personal attitudes. In some colonies, particularly penal colonies such as that in New South Wales, there were some magnificent exceptions such as the failed Governor William Bligh, whose autocratic style derived from his background in naval command. He was ousted in a military coup by a clique of commercial interests.[8] This period of audit at long-distance enabled a degree of discretionary liberty in the British colonies only until about the 1870s, after which the telegraph enabled everyday decision-making to be done from London, and local discretionary powers became rare for colonial functionaries.

Now we turn to technologies of violence. The two hundred years prior to the end of the eighteenth century was a period of generalized violence in Europe, marked by rapid improvements in weapons and fighting. It was a time described by historian Josep Fontana as, "a period of tortures," characterized by, "witch hunts, wars of religion, and inquisitorial persecutions which filled the scientists with fear."[9] Improvements in the uses of gunpowder guaranteed European naval and military superiority over the rest of the planet during the colonial period. Therefore the ultimate sanction was always open to the colonizing institutions through their overwhelming technologies of violence.

Now we turn to technologies of measurement. The calibration of weights and measures is an inheritance from ancient times. For example, here is the bible:

> You shall not have unequal weights in your bag, one heavy, the other light. You shall not have unequal measures in your house, one large, the other small. You shall have true and correct weights and true and correct measures.[10]

Here is the Koran:

> Woe to those who STINT the measure: Who when they take by measure from others, exact the full; But when they mete to them or weigh to them, [di-]minish— What! have they no thought that they shall be raised again, For the great day?[11]

There is a repetitive pattern of legislation and enforcement of standardized weights and measures from ancient civilizations to the present, which reflects the concerns with trade and prosperity (or if you like, scarcity), which are inescapably part of the problem of regulation.

In the economic historian's mirror of hindsight, it is obvious that markets grow really large only where they enjoy the benefit of standardization. By way of illustration, in 1668 the English economist Josiah Child wrote that the great export trade of the Dutch was produced by, "their exact marking of all their Native Commodities; [e.g. hogsheads of fish] buyers will accept of them by the marks, without opening."[12]

Legislate as sovereigns might for uniformity, the actual scales and containers were in the hands of powerful traders. The problem is traceable from early civilizations through to the early-modern period. If we take the problem to be dishonesty as the sources suggest, then we have a straightforward analysis that would appear to fit the facts. Detailed historical investigation reveals, however, that prior to the emergence of late-modern forms of equity value, pressures to vary the measure rather than the price, were not merely the result of dishonesty.

The cause was in the practices of evaluation and the representation. As long as these practices remained stable and dominant, there was nothing in the practices themselves which tended to produce generalized standardization. Therefore during this long proto-capitalist period, legislation and the interests of some groups of traders were the main producers of what standardization was achieved.

I propose now to describe the unpredicted emergence of a generalized market demand for standardized representation. Market-driven standardization has a history which I propose to trace briefly in broad outline. This is important because it explains how it came to be that standardized representation of land emerged historically to remain a dominant feature of late-modern colonial equity value systems of state regulation.

If, prior to the emergence of late-modern equity value systems of state regulation, a landholder wanted to secure a loan using the land as collateral, notarized title deeds manifested the authority of the state with its imputational, compulsive and constraining powers to provide the necessary security and certainty of transaction. The English tradition of legal entitlement to landed estates, like others, required protocols of representation for purposes of state imputation of title. Words like 'premises' and 'cadastral' developed their meanings within these official protocols of imputation. What was on a title deed, in addition to a narrative description of boundaries and a not-to-scale plan, or plat, was a measurement commensurated by traditionally stable engagement between the human and the non-human natural world, rather than the standardized empirical measurement that we now recognise as necessary to equity value systems of state regulation.

The acre under the old commensuration was as an historically stable unit of rent payable in human services. It was not a standard, measured, geometrical area. The emergence of late-modern equity value systems of state regulation came about when a crisis precipitated a change in the way the non-human natural world was represented on title deeds. Title deeds were the instruments of imputation supported by the authority of the state.

A commensuration is a conventional practice. It is a tradition of measurement emerging from experience. In practice, a situation often arises where two things that are measured on incommensurate scales must be taken into account in a repeatable and conventional way so as to permit prediction and audit. It might be as well at this point to consider an illustration of the phenomenon of recommensuration.

In the recent past, surgeons would make decisions about abdominal surgery, giving due weight to cosmetic considerations and to the risk of infection and tended to avoid removing a patient's umbilical. A new commensuration of the two factors was made necessary by the sudden prevalence of savage antibiotic-resistant infections. Since umbilicals are so convoluted they are almost impossible to sterilize, and therefore a potential Trojan-Horse for drug-resistant infections. The new situation is that the reattachment of an umbilical is analogous to catapulting the corpse of a cholera victim into a besieged city. Therefore a new commensuration, a new tradition, has emerged. A living patient without an umbilical is a better result than a corpse with one.

Some surgeons tend now to remove umbilicals. This trend has been forced by the new weighting of factors on the incommensurate scales of cosmetic value and risk management. [13]

Just so, by a new development, sixteenth and seventeenth century relations between the land market and instruments of imputation and compulsion came under a new pressure. A new commensuration arose as a result of that pressure of demand. I propose now first to describe in some detail the earlier commensurative system concerning state imputation and compulsion with regard to land, the pressures brought to bear on that system, and the recommensuration which emerged historically under those pressures. Fred D'Agostino helpfully distinguished the historical process of producing a new tradition of commensuration from the use of an established tradition. [14]

Before Henry VIII, the things measured and commensurated on title deeds (which were the standard imputational instrument concerning land), were twofold. On the one hand was the fecundity of the non-human natural world (which was stable), and on the other, human agricultural practices (which were also stable). An acre varied in size. The size was determined by the ability of the land to support a given number of people in a traditional relationship with the non-human natural world. The fertility and the climate on the one hand and the cropping and husbandry traditions on the other, were each stable and measurable. The commensuration between them produced a tradition of valuation for purposes of taxation that was stable and reliable.

Price fluctuation was no threat to it and the old tradition of commensuration remained stable as long as the acre was a measure of rent payable in human services – which is to say – as long as the things commensurated by the old system remained stable. [15] What was standardized in this paleo-capitalist period, was a socially calibrated assessment of the relation of the human to the non-human natural world and it was done by a locally standardized tradition of commensuration that was stable over time. For example the great survey of England undertaken under William the Conqueror in 1086 measured land according to the richness of the soil. [16] A *virgate* was enough land for a single person to live on and a 'hide' could support a family. Consequently the size shrank when measuring fertile land, and expanded in poor, upland territory. The acre and the *carrucate* were equally variable. An institutional *telos* of dependency on the non-human natural world could remain a feature of governance, only as long as rent was paid in human service. As long as the value of the non-human natural world was tied to the number of humans it could support, then its cadastral area mechanically measured and mathematically calculated, was not used as its measure for purposes of determining its value; and therefore never used for imputation.

In 1538, one event changed things irreversibly, and in so doing triggered the emergence of late-modern equity value systems of state regulation. Late modern equity value systems of governance are extensive formalized systems of commodification of the empirically measured non-human natural world. They express a relation between state agencies of imputation, compulsion and constraint, and private holders of equity. The authenticating imputational authority of a state agency provides certainty to lenders. They thus enable holders of equity to borrow against state-authenticated collateral. This borrowed money enables equity in the non-human natural world to do other commercial work, remote from the actual physical location of the land. Taxes and duties make the relationship symbiotic.

Late-modern equity value systems of state regulation are historically emergent from earlier contingent relationships between states and markets. Late-modern equity value systems of state regulation as I describe them are a symbiosis between equity value and state interests. The symbiosis organizes both, setting up a compulsory modality of relations between humans and towards the non-human natural world. Such systems have no essential nature. They were not born at a single point in history. On the contrary, their roots are spread over millennia and over numerous past civilizations. The reason this essay offers a genealogical investigation of late-modern equity governance is to illustrate that many current institutional practices towards both the human and the non-human natural worlds were produced under pressures of demand in various ways, not least of which was accident.

In 1538 in the greatest land sale in English history before or since, Henry VIII put on the market for cash a half-million acres that had been expropriated from the church. This combined with one other factor to cause an emergency in imputational systems. Lenders demand authentication by title deed, and suddenly title deeds previously adequate were thrown into a crisis of authentication due to destabilization of the thing represented on title deeds. The old acre had been a stable inherited commensuration between the fecundity of the non-human natural world and stable human practices, showing a variable area which could support a fixed number of people.

But now people on the land were being replaced by sheep. It was already evident to the new owners of Henry VIII's seized land, that sheep were more profitable than people. In a section of his 1516 tome, *Utopia* under the heading, "On Enclosures," Thomas More had noted that decent men were turned into thieves by the alienation of common land. Here is the famous passage where he noted that alienated entitlement to land had caused honest men (who were using the land as their ancestors had done from time immemorial) into thieves, and caused sheep to become eaters of people:

> But yet this is not only the necessary cause of stealing. There is another, which, as I suppose, is proper and peculiar to you Englishmen alone. What is that, quoth the Cardinal? forsooth my lord (quoth I), your sheep that were wont to be so meek and tame, and so small eaters, now, as I heard say, be become so great devourers and so wild, that they eat up, and swallow down the very men themselves. They consume, destroy, and devour whole fields, houses, and cities.[17]

By 1538, the old commensuration was already shattered by the alienation of the commons. The consequent displacement of persons destabilized the thing traditionally commensurated, and so set up a demand for a whole new commensuration, a new system of imputation of entitlement to land, a new system of official representation of the non-human natural world. The new commensuration which was to emerge would replace the earlier acre which had been a variable area enabling a fixed price per acre payable in human services. The old acre was replaced in time by the acre as a fixed area, for which a variable cash price was determined by the market.

That an emergency of imputational entitlement for equity value purposes had arisen is illustrated by the fact that in 1538, (the same year as Henry VIII's mass land sale) Sir Richard Benese published a text to standardize empirical measurement by rod or pole precisely sixteen and a half feet long, and to stipulate the drawing of an accurate plan to show shape and extent. There were four square rods in a daywork, and forty dayworks in an acre, and 640 acres in a square mile. [18] Benese's text was a ready reckoner, full of drawings and tables for use in the field. This illustrates the emergence of a demand pressure towards standardized empirical measurement and standardized representation of authenticated imputation.

Nevertheless Benese's text stipulated a different, a longer rod, to measure woodland acres, which were larger than field acres. Thus Benese's work illustrated at the same time the trend to imputation by empirical measure and the inertia of the old complexities; in that it retained a dual system of measurement. Two standard acres remained, one for arable fields and one for woodlands. The trend to accurate, standardized empirical representation which his work illustrates provided a demand pressure which was to eventually sweep his dual system into disuse.

Traditions of approximation in surveying practice were an immediate problem, as Andro Linklater observed. "The surveyor's commonest trick when faced with an irregular shape was to add the lengths of all the sides, divide the total by four, then square the result." The answer thus gotten was quick, easy, and wrong – but usually not by enough to alert the landowner. [19] Edward Warsop wrote in 1582, in *A Discoverie of sundrie errours and faults daily committed by* [surveyors] *ignorant of arithmaticke and geometrie to the damage and prejudice of many of her Majestie's subjects,* "it is the way all Syrveyors do,—whether it originates in idleness, inability or want of sufficient pay, it is not for me to determine." [20] If only the non-human natural world could be measured by a unit that did not vary, supply and demand would determine the price and it could be treated as a standardized commodity. Stephen Johnston expressed the demand pressure:

> Mathematical practice was public: it was made visible and accessible through the printing of vernacular texts on topics such as geometry, arithmetic, algebra, astronomy, surveying, and navigation. Its practitioners emphasized that their work was geared towards active use, rather than conspicuous display or scholarly demonstration and textual correction. These practitioners also claimed that their skills were appropriate to almost any location, whether aboard ship, in the survey of rural or urban land, or on the battlefield. [21]

A demand that the non-human natural world be represented with cadastral accuracy had begun. A market-driven demand had begun for accurate and auditable standardized technologies of survey. The historic significance of the emergence of this demand is that because it was a condition of the standardized representation of imputation of equity value, it would become a permanent feature of equity value systems of state regulation. Prior to this time standardization was a trade-friendly top-down enterprise. From this time, market-driven (bottom-up) standardization, has remained a permanent and familiar dynamic.

A Belgian named Simon Stevin first published an account of decimals in 1585 and an Englishman, one Edmund Gunter, was quick to realize that working in decimals rather than fractions was extremely effective in logarithmic and trigonometric calculation for survey purposes. [22] Decimals, logarithms, trigonometry and geometry had been generally available only in Greek and Latin prior to Gunter's 1623 publication, *The Description and Use of the Sector, the Crosse-staffe, and Other Instruments for Such as are Studious of Mathematical Practise.* [23]

Gunter invented a reliable and uncomplicated ground measurement device that enabled the emergence of late-modern equity value systems of governance. His device was a commensuration. It married the English measurement tradition, which was in multiples of four, with the convenience of the newly imported decimal system. Gunter's device was a chain (four rods) long, divided into a hundred links with brass links grouped in tens. This had the effect of creating a new commensuration of the two systems of measurement, overlaying the convenience of the new decimal scale on the original English scale of measurement by multiples of four. Ten square chains measured precisely one acre. The genius of Gunter's device was that it forced even the least competent surveyors to standards of exactness, and enabled their work to be reliably auditable. Taken together, the circumstance of a sudden large market in land and the consequent invention of Gunter's chain can be seen in retrospect as necessary causal elements in the development of the mass commodification of entitlement to land.

Whether Edmund Gunter in his short life realized it or not, what he had devised was a means of quickly generating equity value in the as yet unsurveyed colonies. It remained to develop the market and reliable symbiotic regulatory arrangements in such a way that complete strangers might reliably and safely enter into a commercial relationship with the non-human natural world conceived as land, without ever even seeing the land.

The record of an instrument of imputation of entitlement, which together with previous records forms a *chain of title*, unlocks the potential of the entitlement to land as alienable equity value. The land entitlement thus certified, takes on a second property. That is, an imputed property – the property of equity value – which can be borrowed against, in order to do commercial work beyond the capacity of the land itself. [24] Institutionally enforced mechanisms of transfer were ancient, but subsequently arrangements were developed which included certified survey plans and records of mortgage, lease, and other relevant information in standardized formats.

Adapted state institutional arrangements (new traditions) emerged due to pressure from commercial interests, to enable complete strangers to do business in commercial safety. [25] Institutionally-produced commercially safe transactions between strangers – including corporate entities separated by distance – became a key requirement for state institutions producing and guaranteeing *equity value*. Hernando de Soto has theorised this as the *sine qua non* of late-modern equity value systems of governance. [26] By way of the developing symbiosis between instruments of state registration and private equity, technologies of adjudication and insurance were adapted to insure all parties against the inevitable occasional mistake. Once these symbiotic technologies of governance and private equity developed out of contingent historical events and institutional adaptations, the institutional meanings of British colonial sovereignty were set on a course that had important implications for the human and the non-human natural worlds.

Now we turn our attention to technologies of finance. Enough of the forms of paleo-capitalism are traceable from the ancient world for us to trace the history of some of the forms of capitalism from then. Yet until the end of the seventeenth century, equity value systems developed mostly as a result of equity value in land. Land rather than securities was the basis of most investment before 1600. [27] On the account of securities historian Ranald Michie:

> It was only from the seventeenth century onwards, with the appearance of both negotiable instruments representing national indebtedness and transferable stocks issued by corporate enterprise, that the volume of business generated by securities was such as to justify the beginnings of professional intermediation and organized markets. [28]

During the seventeenth and eighteenth centuries, technologies of finance were developed, particularly in Amsterdam which was the centre of the world trade in securities. These technologies were to become important to British colonial systems of equity value creation. [29] Although the history of financial instruments extends back to the earliest civilizations, the Dutch invented new instruments for trade in securities, such as time bargains, price lists, and dealing for the account.[30] The effectiveness of these new technologies developed in the sixteenth century depended on certain political controls becoming stabilized. In England the interests of the landed aristocracy aligned with those of the wealthy middle-class to assume political control of the state *via* constitutional monarchy established through the bloodless coup of 1688. The permanent removal of arbitrary Crown interference made government debt a secure commodity. This enabled an explosion of speculative trade in stocks (government debt).[31] Speculation in government debt made many fortunes, and by the turn of the century there were quite a few illiterate Cockney millionaires living like lords, celebrating new forms of social mobility. Fortunes were also lost in less secure speculations; bubbles, as they came to be called.

In contrast with the autocratic European monarchies such as the French and Spanish, which could not borrow except at high rates of interest, in England government debt came to be bought and sold by means of the newly invented financial instruments. England was now able to finance immensely costly projects such as prolonged wars and colonial enterprises. In the words of Ranald Michie: "The very size of national debts created problems of matching buyers and sellers that could only be solved through organized markets."[32]

By the end of the eighteenth century, Paris and Amsterdam financiers had folded their tents and moved to London, which was thereafter for the colonial period, the financial capital of the world. Because colonial enterprises required such heavy government expenditure, the British were uniquely placed by financial technologies, without which other European powers were uncompetitive.[33] A new era ensued, in which there was on offer a hyphenation of imputational instruments; one kind of imputational instrument for land equity value, and another for short-term speculation in other forms of equity value. Both systems of entitlement went hand-in-glove; a symbiosis enabling the strategic mixing of long-term and short-term financial options.

This brief survey of millennia of financial technologies has skipped a number of important developments such as the normalization of interest and double-entry accounting, to mention only two. What matters is that timely adaptations of financial technologies both enabled and drove speculation, which in turn enabled previously unimagined levels of government debt. Those massive government debts enabled success in war and the domination and expropriation of colonized peoples around the entire planet. Combined with the other technologies listed here, this trade in government debt also enabled the symbiotic commodification of entitlement to land in the remotest reaches of the planet.

Indigenous Subordination

I contend that relations between the human and the non-human natural worlds are intelligible *via* the auditable institutions and procedures of the time. I challenge the assumption that British values and attitudes are to be best found by careful examination of individual values and attitudes in the texts of the period.

A civil *telos* tends to be overlooked by the blindly individualistic when it comes to moral life. The energizing *telos* of a civilization has a life of its own, independent of individual defection, that is, independent of individual consciousness. Every one of us is familiar with the phrases like, "that's the way it is," and, "I'd like to help you but it's out of my hands." A civilizational *telos* is analogous to the Abilene Paradox in which nobody actually wanted to go to Abilene, but everybody thought everybody else did. If you don't do what's expected, you know that someone will replace you who will; you have no real choice. That's what a civil *telos* is. It's why the last tree on Easter Island was cut down.[34]

The British colonial *telos* produced institutional technologies of domination and appropriation. British colonial treatment of the non-human natural world was produced by the emergence of late-modern equity value regulatory systems, which in turn created new possibilities of thinking about the problems of social order. The new shapes of social order produced by emergent capitalism also created the conditions for the production of new challenges to old ways of thinking about social order.

The new social division was between those who might hope to acquire equity value and those who might not. The colonizers were also divided in the same way; by the surveying and conversion of the non-human natural world to equity held by some and not others; and the production of social order consequent on the operations of this social division. You were rich if you had money in your pocket on a Saturday night in the colonies, but if you were wealthy you had your name entered on a certified plan – a title deed.

The new social division was bred from the fusion of two cultural products. The first product is equity value, a product of alienation. The archaic English meaning of *alienation* indicated improper use, but by the late middle ages it came to mean lawful transfer of ownership. Equity value is produced by the convention that a person can have a connection with the non-human natural world by an act of institutional imputation, rather than *via* a lived dependency relationship.

All value is produced by imputation and compulsion. We love our children, we desire their safety. To desire their safety is not the same thing as placing a value on it. To value their safety, to place a value on their safety, is to make and pay for arrangements to keep them safe. Their safety is not a value unless it is manifest in arrangements to keep them safe. The social dimensions of such arrangements will always include the placing of an obligation of constraint on those who might profit from or enjoy endangering our children, as well as the production of appropriate incentives and disincentives, and the advertising of them. In stable societies institutions can be relied upon to value our children through the operation of imputation, compulsion and constraint. In the case of the value of our children, the imputation is real, the compulsion is real, the constraint is real, the value is real, and the children are real. The entitlement created by the imputation and compulsion operates to provide certainty to real people about the safety of real children. When we come to equity value in land, we find a different story. The entitlement created by imputation and compulsion operates to provide certainty to real people about a cultural product. Alienation of land is a notional connection between real people (or sometimes other legal entities such as corporations which are considered artificial persons) and real non-human natural environments, backed by the authority of the state. This cultural product is created at the point of imputation, at the intersection between the polity and the putative holder of equity value.[35] The entitlement thus created has a money value; a money value exactly equal to the amount a lender will lend on the strength of the certainty provided by the instruments of imputation and compulsion.

Such immense power and prosperity emerged from the legal imputation of equity value through the technologies just described, that imputation of equity became a major engine of governance. Karl Marx wrote, "Estrangement is manifested in the fact that *my* means of life belong *to someone else*...but also in the fact that everything is itself something different from itself."[36] Marx emphasized the first part in his *labor theory of value*. The second part, 'everything is itself something different from itself' has proved in hindsight to have been the real engine of capitalism, rather than labor. The treatment of the non-human natural world as, "itself something different from itself," by regulatory imputation has proved in hindsight to have been more fundamental to the success of equity value than surplus labor ever was. The regulatory imputation of alienation, giving rise to entitlement with its attendant compulsions and constraints on other humans was the source of the social division between holders of equity and non-holders.

The second cultural product is the credit note, or merchant's bill of exchange. By the year 1700 CE, credit instruments are estimated to have exceeded the coinage of England in money value, by a quarter.[37] Credit depends for its value on an act of belief, (hence the word), and not on any necessary connection with the non-human natural world. Credit has no necessary relation to the value of the non-human natural world, and no usefulness except in markets – it is inconstant, changing upon opinion and passion, and beyond control. This cultural product emerged historically alongside official instruments of imputation, and the resultant hybrid arrangements forced the equity split which was the dominant feature of British sovereignty for indigenous peoples.

Now that the actual social order was so transparently a product of the operations of equity value governance rather than of traditional dependency on the non-human natural world, a new market demand emerged for the re-theorizing of the relations of prosperity, of meaning, and of social order.[38] The old explanatory systems were cast aside by theorists such as Voltaire, Tom Paine, Thomas Jefferson, Jeremy Bentham, Adam Smith and Thomas Malthus, in favor of new models of market-produced citizenship for a hungry market in ideas.[39] The challenge to *ancien regime* systems of meaning was also manifest in the ease with which old local arrangements gave in to standardizing practices such as regulatory centralization and statistical measurement. Napoleon Bonaparte introduced centralized standardized regulation in one fell swoop. Centralization took a century longer in Britain, but its time had come.[40]

Operational relationships between colonizers and colonized were constrained by the practicalities of establishing connections between their two worlds, their worlds of customary practice. Since we have tracked only the institutional practices of the colonizers, it remains to consider what possibilities were opened within British regulatory practices for treatment of the colonized. The first is the meaning of the imputation of exclusive entitlement to the equity holder. This entitlement is produced in part by the enactment of sanctions constraining generalized obligations of forbearance. These obligations of forbearance which in part constitute entitlement are official. State institutions produce and enforce subordination, subservience, cooperation, acquiescence, and compliance. The entire histories of European colonial enterprise are marked by energetic institutional efforts at producing such constraints for colonized peoples.

The second locus of value traceable in British colonial institutions is the value expressed by the market in terms of equity. This logic of the market leaves an operational gap, consisting in all that is neither owner nor owned, into which all persons who are not owners fall. The market in land (entitlement to control of the non-human natural world commodified and expressed as equity value) was not, at the end of the eighteenth century, institutionally open to indigenous peoples in most of the British colonies and in the emergent thirteen United States of America. That is to say, by and large no provision was made, no procedures created, no arrangements produced, to enable them, if they were capable of wishing it, to buy back control of their own land by means of having it surveyed and registered. This was because they were not institutionally categorized as participants in the market in land entitlements expressed as equity value. There were exceptions such as in New Zealand. Like most of the British themselves, indigenous people were rarely offered a seat at the table of equity value. They were not in any direct sense beneficiaries of the dominant meaning of British governance. That dominant meaning was expressed by fences and gunpowder, and by hyphenations of enslavement. British colonies had a different historical trajectory from the Spanish who vested colonial land in the name of the crown. It remained to the colonized indigenous peoples in the British colonial tradition, to subordinate, to cooperate, to acquiesce, and to comply with the obligations arising from the imputation of their own ancient entitlements to strangers.

Thus the meanings of British colonial governance at the end of the eighteenth century were manifest in the alienability of the non-human natural world. Some indigenous groups were treated as having no more legal standing than fauna. Equity value was constructed *via* a state-authenticated market. This institutionally created product, this sudden creation of new equity value (new entitlements created as fast as survey parties could cover the country), this creation of new money by new title deeds, derived from the British institutional past. It separated the meaning systems of the colonizers and the colonized, the occupiers and the occupied. The institutional membrane that separated those who valued entitlement to the non-human natural world *via* a market and those whose practices towards the non-human natural world were not primarily market-mediated, was not merely conceptual. At the level of everyday practice, the sovereignty of the state changed indigenous relationships with the land by framing the shapes of public compulsion and constraint to produce exile and in some cases quasi-enslavement.

My purpose here is not to make moral comparisons. This essay attempts no description of indigenous meaning systems for comparative purposes. Those meaning systems may well have been as artificial and contingent, or as cruel and heartless, as any other, I cannot know. This genealogical framing of late-modern equity value regulatory systems has no moral purpose, but rather serves as a pointer to the conceptual work that regulators have ahead of them when it comes to redesign of state procedures.

This section has had two purposes. The first has been to establish the historical contingency of the technologies of commensurative tradition between the non-human and the human worlds *via* a targeted historical survey. We have seen the roles played by accident in the emergence and adoption of convenient, powerful cultural products. It is worth pondering that the shapes of entitlement which million now take for granted might have developed otherwise. Secondly I have sought to establish that at the end of the eighteenth century, peoples colonized by the British were subjected to sanctioned compulsions and constraints institutionally imposed, consequent to the meanings expressed in British audited institutions. The level of historical generalization at which this case has been made is so broad as to bridge many particular exceptions and to omit many qualifying factors. I offer two examples: the penal roles of some colonies and the practices of leasing Crown land in many instances seriously punctuated colonial development of freehold entitlement to land as equity value.

This brief demonstration that there was an institutionally expressed British colonial *telos,* could be extended to investigate in more historical detail the axiologies and epistemologies variously expressed in the audited procedures of various British colonial regulatory institutions. It might be helpful for example, to consider notions of agency expressed within the colonizing tradition. [41] It is predictable that such a discourse of investigation would be a field of contestation between historians and that it would uncover a far from uniform field of practices. Apart from illustrating to some degree the institutional conditions under which colonised peoples found themselves under British rule at the end of the eighteenth century, the purpose of this section has been to illustrate three historical observations, as follows.

Hybridizations: The Making of New Traditions

New traditions of commensuration have been thrown up historically by a variety of causes, some invented, some accidental, and some evolutionary. A new tradition is sometimes produced as an unintended consequence, such as that arising from Henry VIII's annexation of monastery lands. This was historically commensurative in that it permanently rejigged the way that the value of the non-human natural world was measured. This rejigging, forced by the historical surprise of a mass land sale and its attendant destabilizations, had results that none of the protagonists involved could have theorized. None of the regulators who had legislated the standardization of weights and measures in many civilizations over several millennia, is remembered for predicting that a mass land sale might, given the right conditions, permanently put in place the 'bottom-up' market pressure for standardization which has been a feature of markets to some degree ever since. Therefore it may be helpful to think of some instances of emergent commensuration as analogous to mutation in evolution, in which the failure rate may be so high in the context of new pressures, that the possibilities of a whole new emergent equilibrium of conventions of commensuration may be impossible to predict.

The second historically emergent commensuration was of land measurement *via* Gunter's adaptation, Gunter's invention. His chain replaced the use of wooden rods for land measurement, but his groundbreaking commensuration idea was the possibility of quick and accurate calculation using decimals. As a result of destabilization of the representation of entitlement to land, replacing measurement in human services by empirical measurement, a new market-driven demand to standardize cadastral measurement developed. This demand resulted in the translation of logarithmic and trigonometric works into English and in Gunter's chain. The technology produced what Gunter (adaptive genius though he was) could not have planned, the future colonial mass market in land as equity value.

Thirdly, recommensurating the incommensurate in order to establish a new tradition can also be a regulatory undertaking performed under conditions of changed necessity and changed pressure. Under these conditions regulators have, from time to time, understood philosophical principles as sources of problems rather than sources of answers. The recommensurative task tends to be seen by regulators as operational rather than conceptual. Regulators make experimental adaptations, often insufficiently theorized, in part purposely left vague for the sake of avoiding the complications of ideologically driven disagreements. [42]

Regulatory adaptation is a process in which there are complex institutional forces at play. For example, conditions of audit are a regulatory given, and conditions of equilibrium are much more readily audited than transitional and experimental arrangements. Experimental arrangements such as those in which redesigned procedures are tested, tend to require personal judgments and discretionary powers which resist conditions of audit and of reliable prediction normal to regulatory systems. Therefore regulatory systems have a tendency, like those decorative glass birds with a reservoir of colored water in their bodies which tilt to drink, a tendency to stabilize to a new equilibrium.

British regulatory systems did generate a new tradition of commensuration to meet the needs of an emergent mass market in land as equity value. The emergent demands of equity forced a new regulatory tradition of guaranteeing equity value entitlements *via* alienation of the non-human natural world.

The task of creating a new commensurative tradition was a task of piecemeal adjustment of procedures and checking for unintended consequences rather than wholesale application of principle. In fits and starts the new demands of land alienation called forth institutional improvements in equity value transfer. The demands of a mass market in land led to the eventual state registration of certified survey plans and records of mortgages, liens, and leases in a standardized audited format. These provisions needed to be made to work experimentally, while maintaining the legal integrity of the old scale of land tenure. During a transitional period the implications for public order and confidence in regulatory institutions had to be managed.

Historically the new state procedures were introduced under commercial pressure to enable complete strangers to do business at a distance with commercial safety. The implementation of new procedures took place while the old system was still running. To appreciate the degree of skill required in managing such a seriously great new regulatory commensuration, consider the pressures of gain, cost, and risk experienced by holders of equity and reflected in the pressures of demand on regulators.

Indigenous Claims and the Redesign of Procedures

One might be tempted to think that if a group of indigenous citizens is able to establish by way of justification that some of their traditional conventions of programmed engagement with the non-human natural world still deserve their loyalty as a form of the good life, then a regulator in a multicultural polity might seek at the very least to offer them better procedures than the British Crown did for their ancestors who were not citizens. This essay has established that British colonial regulators treated both human and non-human natural worlds as subordinate to the interests of holders of equity value.

Nevertheless, one might be forgiven for imagining that in the centuries since first colonization, indigenous peoples and colonizers must have developed some synchretized accommodations, some *modus vivendi,* some hybrid emergent traditions. Even though accommodations can often develop without being theorized, I recommend the active cultivation of such accommodations in the redesign of procedures for the hearing of indigenous land claims. The root of justification for this is that the work of governance is to ensure the peace and prosperity of the polity. Just as natural diversity increases the resilience of an ecosystem, cultural difference can, in some cases, increase the resilience of a civilization. Thomas Homer-Dixon generalized that highly adaptive complex systems tend to have diverse constituent elements, decentralized power, and open sharing of information. Such systems also tend to be, "unstable enough to create unexpected innovations, orderly enough to learn from failure and success." Such systems, "stimulate constant experimentation, and generate a variety of problem-solving strategies."[43]

Indigenous engagements with the non-human natural world are sources of intellectual wealth which regulators might take seriously if they are to pursue the long-term interests of the state. Therefore regulators in multicultural states should adapt their procedures by engagement, negotiation, and accommodation to enable and preserve the intellectual and cultural wealth of indigenous groups in the interests of the future prosperity of the polity.

Bibliography

Cardoza, Rod, "Evolution of the Sextant," in *Nautical Brass Magazine Online,* Web: home.earthlink.net/~nbrass1/cardart.htm Accessed: 26 January 2008.

Chadwick, Edwin, *The Evils of Disunity in Central and Local Administration: Especially with Relation to the Metropolis: and also on the New Centralization for the People together with Improvements in Codification and in Legislative Procedure,* (London: Longmans, Green, 1885).

Chancellor, Edward, *Devil Take the Hindmost: A History of Financial Speculation,* (London: Papermac, 2000).

Cockburn, J.S., King, H.P.F., McDonnell, K.G.T. (eds.), *A History of the County of Middlesex: Volume 1: Physique, Archaeology, Domesday, Ecclesiastical Organization, The Jews, Religious Houses, Education of Working Classes to 1870, Private Education from Sixteenth Century* (London: Victoria County History 1969). pp. 90-94. Web: www.british-history.ac.uk/report.aspx?compid=22105. Accessed: 31 January 2008.

Cohen, Felix S., "Dialogue on Private Property," Rutgers Law Review, Vol. 9.

Cotter, Charles H., *A History of the Navigator's Sextant,* (Glasgow: Brown Son & Ferguson, 1983).

Cotter, Charles, "Edmund Gunter (1581-1626)," Journal of Navigation, Vol. 34. No. 3. 1981.

D'Agostino, Fred, *Incommensurability and Commensuration: The Common Denominator,* (Aldershot: Ashgate, 2003).

de Soto, Hernando, *The Mystery of Capital: Why Capitalism Triumphs in the West and Fails Everywhere Else,* (New York: Basic Books, 2000).

Diamond, Jared, *Guns, Germs and Steel: The Fates of Human Societies* (New York: W. W. Norton &Company, 1997).

Dijksterhuis, E. J., et al. *The Principal Works of Simon Stevin,* (Lisse: Swets & Zeitlinger, 1955-1966).

Fontana, Josep, *The Distorted Past: A Reinterpretation of Europe,* [translated by Colin Smith], (Oxford: Blackwell, 1995).

Foucault, Michel, *Discipline and Punish: the Birth of the Prison,* [translated by Alan Sheridan], (New York: Pantheon, 1977).

Gould, Rupert, *The Marine Chronometer*, (London: The Holland Press, 1960).

Gray, Kevin, "Property in Thin Air," The Cambridge Law Journal, Vol. 50. 1954-55.

Harvey, Jerry B., "The Abilene Paradox: The Management of Agreement," Organizational Dynamics, Vol. 17. Issue 1. Summer 1988, pp. 17–43.

Holling, C. S., "Understanding the Complexity of Economic, Ecological, and Social Systems," Ecosystems, Vol. 4. No. 5. 2001.

Homer-Dixon, Thomas, *The Upside of Down (Catastrophe, Creativity, and the Renewal of Civilization),* (Melbourne: Text publishing, 2006).

Johnston, Stephen, "The identity of the mathematical practitioner in 16th-century England," in Hantsche, Irmgarde, (ed.), *Der "mathematicus:" Zur Entwicklung und Bedeutung einer neuen*

Berufsgruppe in der Zeit Gerhard Mercators, Duisburger Mercator-Studien, Volume 4, (Bochum: Brockmeyer, 1996).

Linklater, Andro, *Measuring America: How the United States was Shaped by the Greatest Land Sale in History,* (London: HarperCollins, 2002).

Mabo v. Queensland (No. 2) (1992) 107 ALR (HC).

MacIntyre, Alasdair, *Whose Justice, Which Rationality?* (Indiana: University of Notre Dame Press, 1988).

Malone, Dumas, *Jefferson and the Rights of Man,* (Massachusetts: Little, Brown, 1951).

Marx, Karl, *Capital Volume III,* (Moscow: Progress Publishers, 1959).

McGuinness, Bruce, and Walker, Dennis, "Sharing History – Recognising Aboriginal and Torres Strait Islander Pioneers," Indigenous Law Resources, Reconciliation and Social Justice Library, 1985.

Michie, Ranald, *The London Stock Exchange,* (Oxford: Oxford University Press 2001).

More, Thomas, *Utopia,* Book 1 (1516); cited in Craik, Henry, (ed.) *English Prose. Volume I,* (New York: The Macmillan Company, 1916).

Morphy, Frances, and Morphy, Howard, "The Blue Mud Bay Case: Refractions through Saltwater Country," Dialogue, Vol. 28. No. 1.

Needham, Joseph, *Science and Civilization in China Volume 1,* (Cambridge: Cambridge University Press, 1954).

Nietzsche, Friedrich, *The Genealogy of Morals*, [translated and edited by Douglas Smith], (Oxford: Oxford World's Classics, 1887 [1996]).

Paddock, W.C., "Our Last Chance to Win the War on Hunger," Advances in Plant Pathology, Vol. 8. 1992.

Paine, Tom, *The Selected Work of Tom Paine and Citizen Tom Paine,* [edited by Howard Fast], (New York: Random House, 1945).

Plantagenet, King John I, *Magna Charta,* clause 13, [translated by the British Museum], in Marsh, Henry, (ed.) *Documents of Liberty: from Earliest Times to Universal Suffrage,* (Newton Abbot: David and Charles, 1971).

Plumwood, Val, "The Concept of a Cultural Landscape," Ethics and the Environment, Vol. 11. No. 2. 2006.

Scott, W. R., *The Constitution and Finance of English, Scottish and Irish Joint-Stock Companies to 1720,* (Cambridge: Cambridge University Press, 1910).

Scull, Andrew, *Madhouses, Mad-doctors, and Madmen,* (London: Athlone, 1981).

Smith, Adam, *An Inquiry into the Nature and Causes of the Wealth of Nations,* (London: Methuen, 1776 [1904]).

Sobel, Dava, *Longitude: the True Story of a Lone Genius who Solved the Greatest Scientific Problem of his Time,* (New York: Penguin, 1996).

Spigelman, James Jacob, *Australia Day Address,* (Sydney: 22 January 2008).

Sunstein, Cass R., *Legal Reasoning and Political Conflict,* (Oxford: Oxford University Press, 1996).

Sunstein, Cass R., *Why Societies Need Dissent,* (Massachusetts: Harvard University Press, 2003).

Tainter, Joseph A., *The Collapse of Complex Societies,* (Cambridge: Cambridge University Press, 1988).

Tan, Poh-Ling, Webb, Eileen, and Wright, David, *Land Law, 2nd edition,* (Australia: LexisNexis Butterworths, 2002).

The Koran, [translated by J.M. Rodwell], (1876).

The New English Bible with Apocrypha (Great Britain: Oxford University Press & Cambridge University Press, 1961 [1970]).

Thucydides, *History of the Peloponnesian War*, [translated by Warner, Rex], (Middlesex: Penguin, 1954).

Tittler, Robert, "Late Medieval Urban Prosperity," The Economic History Review, New Series, Vol. 37. No. 4. 1984.

Torrence, John, *Estrangement, Alienation and Exploitation*, (London: The Macmillan Press, 1977).

Voltaire, *The Complete Works of Voltaire* [edited by Theodore Besterman et al.] (Oxfordshire: Voltaire Foundation, 1968).

Webber, Jeremy, "Beyond Regret: Mabo's Implications for Australian Constitutionalism," in Ivison, Duncan, Patton, Paul, and Sanders, Will, (eds.), *Political Theory and the Rights of Indigenous Peoples*, (Cambridge: Cambridge University Press, 2000).
Williams, Bernard, *Truth & Truthfulness: an Essay in Genealogy*, (Princeton: Princeton University Press, 2002).

Endnotes

[1] Thomas More, *Utopia*, Book 1 (1516); cited in Henry Craik (ed.), *English Prose. Volume I*, (New York: The Macmillan Company, 1916). pp. 162-164.

[2] Hernando de Soto, *The Mystery of Capital: Why Capitalism Triumphs in the West and Fails Everywhere Else*, (New York: Basic Books, 2000); Andro Linklater, *Measuring America: How the United States was Shaped by the Greatest Land Sale in History*, (London: HarperCollins, 2002); Fred D'Agostino, *Incommensurability and Commensuration: The Common Denominator*, (Aldershot: Ashgate, 2003); and Paul Ivory, Personal Correspondence, (23 April, 2007).

[3] Mabo v. Queensland (No. 2) (1992) 107 ALR (HC). For its implications see, for example: Jeremy Webber, "Beyond Regret: Mabo's Implications for Australian Constitutionalism," in

Duncan Ivison, Paul Patton and Will Sanders (eds.), *Political Theory and the Rights of Indigenous Peoples*, (Cambridge: Cambridge University Press, 2000). pp. 60-88.

[4] Thucydides, *History of the Peloponnesian War*, [translated by Rex Warner], (Middlesex: Penguin, 1954). p. 402. The Melian Dialogue.

[5] Robert Tittler, "Late Medieval Urban Prosperity," The Economic History Review, New Series, Vol.37, No.4. November 1984. pp. 551-554. See also King John I (Plantagenet), *Magna Charta*, clause 13, [translated by the British Museum], in Henry Marsh, (ed.) *Documents of Liberty: from Earliest Times to Universal Suffrage,* (Newton Abbot: David and Charles, 1971). p. 43.

[6] Joseph Needham, *Science and Civilization in China Volume 1,* (Cambridge: Cambridge University Press, 1954). p. 242.

[7] Rod Cardoza, "Evolution of the Sextant," in *Nautical Brass Magazine Online,* Web: home.earthlink.net/~nbrass1/cardart.htm Accessed: 26 January 2008. The Hadley Quadrant, the sextant, and the chronometer came into general use in the late eighteenth century. See also: Charles H. Cotter, *A History of the Navigator's Sextant,* (Glasgow: Brown Son & Ferguson, 1983); Rupert Gould, *The Marine Chronometer,* (London: The Holland Press, 1960); and Dava Sobel, *Longitude: the True Story of a Lone Genius who Solved the Greatest Scientific Problem of his Time,* (New York: Penguin, 1996).

[8] James Jacob Spigelman, *Australia Day Address,* (Sydney: 22 January 2008). The Hon. James Spigelman is (in 2010 CE) the Chief Justice of New South Wales.

[9] Josep Fontana, *The Distorted Past: A Reinterpretation of Europe* [translated by Colin Smith], (Oxford: Blackwell, 1995). p. 162. See also Jared Diamond, *Guns, Germs and Steel: The Fates of Human Societies,* (New York: W. W. Norton & Company, 1997).

[10] Deuteronomy 25:13-15, *The New English Bible with Apocrypha*, (Great Britain: Oxford University Press & Cambridge University Press, 1961 [1970]).

[11] *The Koran*, Sura 83:1-5, [translated by J.M. Rodwell], (1876).

[12] Andro Linklater, *Measuring America: How the United States was Shaped by the Greatest Land Sale in History*, (London: HarperCollins, 2002) pp. 16-22.

[13] In 2008, a Russian surgeon almost killed the author by re-attaching his umbilical, which carried a multi-drug resistant infection.

[14] Fred D'Agostino, *Incommensurability and Commensuration: The Common Denominator*, (Aldershot: Ashgate, 2003). p. 4.

[15] Andro Linklater, *Measuring America: How the United States was Shaped by the Greatest Land Sale in History.* (London: HarperCollins, 2002) pp. 5-6.

[16] J.S. Cockburn, H.P.F. King, K.G.T. McDonnell (eds.), *A History of the County of Middlesex: Volume 1: Physique, Archaeology, Domesday, Ecclesiastical Organization, The Jews, Religious Houses, Education of Working Classes to 1870, Private Education from Sixteenth Century*, (London: Victoria County History 1969). pp. 90-94. Web: www.british-history.ac.uk/report.aspx?compid=22105. Accessed: 31 January 2008.

[17] Excerpt from Thomas More, *Utopia*, Book 1 (1516); cited in Henry Craik (ed.) *English Prose. Volume I* (New York: The Macmillan Company, 1916). pp. 162-164.

[18] Richard Benese, *The boke of measuryng of lande as well of woodland as plowland, [and] pasture in the feelde: [and] to compt the true nombre of acres of the same. Newly corrected, [and] compiled by Sir Richarde de Benese*, [Second Edition

edited by Thomas Paynell] (London: Thomas Colwell, [1563]). pp. 17-18.

[19] Andro Linklater, *Measuring America: How the United States was Shaped by the Greatest Land Sale in History,* (London: HarperCollins, 2002). p. 13.

[20] Andro Linklater, *Measuring America: How the United States was Shaped by the Greatest Land Sale in History,* (London: HarperCollins, 2002). p. 13.

[21] Stephen Johnston, "The identity of the mathematical practitioner in 16th-century England," in Irmgarde Hantsche (ed.), *Der "mathematicus:" Zur Entwicklung und Bedeutung einer neuen Berufsgruppe in der Zeit Gerhard Mercators, Duisburger Mercator-Studien, Volume 4,* (Bochum: Brockmeyer, 1996). pp. 103-104.

[22] Virtually all of Stevin's writings have been published in five volumes with introduction and analysis in E. J. Dijksterhuis, et al. *The Principal Works of Simon Stevin,* (Lisse: Swets & Zeitlinger, 1955-1966).

[23] See Charles Cotter, "Edmund Gunter (1581-1626)," Journal of Navigation, Vol. 34. No. 3. 1981. pp. 363-365.

[24] Karl Marx, *Capital, Volume III,* (Moscow: Progress Publishers, 1959). p.314. "Estrangement (alienation) is manifested in the fact that *my* means of life belong *to someone else*...but also in the fact that everything is itself something different from itself."

[25] For example, Robert Torrens introduced such regulatory reforms in Australia. See Poh-Ling Tan, Eileen Webb, and David Wright, *Land Law 2nd edition,* (Australia: LexisNexis Butterworths, 2002). p. 170.

[26] Hernando de Soto, *The Mystery of Capital: Why Capitalism Triumphs in the West and Fails Everywhere Else,* (New York: Basic Books, 2000).

[27] W. R. Scott, *The Constitution and Finance of English, Scottish and Irish Joint-Stock Companies to 1720*, (Cambridge: Cambridge University Press, 1910). pp. 44, 155, 161.

[28] Ranald Michie, *The London Stock Exchange*, (Oxford: Oxford University Press 2001). p. 2.

[29] Ranald Michie, *The London Stock Exchange*, (Oxford: Oxford University Press 2001). p. 3.

[30] Edward Chancellor, *Devil Take the Hindmost: A History of Financial Speculation*, (London: Papermac, 2000). p. 32. These instruments were called Dutch finance.

[31] Edward Chancellor, *Devil Take the Hindmost: A History of Financial Speculation*, (London: Papermac, 2000). p. 32. Parliament underwrote all government debt in 1693, and the Promissory Notes Act of 1704 made all debts transferable.

[32] Ranald Michie, *The London Stock Exchange*, (Oxford: Oxford University Press 2001). p. 2.

[33] See: Edward Chancellor, *Devil Take the Hindmost: A History of Financial Speculation*, (London: Papermac, 2000). p. 32.

[34] Historiography that does not rest on the primacy of the consciousness of individual subjects is not new; a form of it metaphorically entitled "archeology" was attempted by Michel Foucault with variable success. This historiographical approach cannot of itself prevent historians looking in the wrong places. I attempt here a version of this historiography that owes more to Jared Diamond and Joseph Tainter, than to Michel Foucault. See Michel Foucault, *The Archaeology of Knowledge*, [translated by A.M. Sheridan Smith], (London: Tavistock, 1972), Jared Diamond, *Guns, Germs and Steel: The Fates of Human Societies*, (New York: W. W. Norton & Company, 1997), Joseph Tainter, *The Collapse of Complex Societies*, (Cambridge: Cambridge University Press, 1988).

[35] Kevin Gray, "Property in Thin Air," The Cambridge Law Journal, Vol. 50. p. 252. Gray's study of property underlines its status as a cultural product: "the ultimate fact about property is that it does not really exist: it is mere illusion." I both agree and disagree with Gray; entitlement is a mere institutional construct, but if it is defined by its institutional, social and environmental effects, then it is no illusion.

[36] Karl Marx, *Capital, Volume III*, (Moscow: Progress Publishers, 1959). p. 314.

[37] Edward Chancellor, *Devil Take the Hindmost: A History of Financial Speculation*, (London: Papermac, 2000). p. 32.

[38] Paul Ivory, Personal Correspondence, (23 April, 2007).

[39] Voltaire, *The Complete Works of Voltaire* [edited by Theodore Besterman et al.] (Oxfordshire: Voltaire Foundation, 1968). Tom Paine, *The Selected Work of Tom Paine and Citizen Tom Paine,* [edited by Howard Fast], (New York: Random House, 1945). Dumas Malone, *Jefferson and the Rights of Man,* (Massachusetts: Little, Brown, 1951). Jeremy Bentham, "Political and Quasi-political Fictitious Entities," in John Bowring, (ed.), *The Works of Jeremy Bentham*, (Edinburgh: William Tait, 1843). p. 205. Adam Smith, *An Inquiry into the Nature and Causes of the Wealth of Nations*, (London: Methuen, 1776 [1904]). Thomas Malthus theorized population growth as a, "cancerous tumor" cited in W.C. Paddock, "Our Last Chance to Win the War on Hunger," Advances in Plant Pathology, Vol. 8. 1992. p. 202. Josiah Wedgwood famously expressed the new need of equity value regulatory systems for a reformed work force, in the hope to, "make such machines of men as cannot err." Andrew Scull, "Moral Treatment Reconsidered: Some Sociological Comments on an Episode in the History of British Psychiatry," in Andrew Scull, *Madhouses, Mad-doctors, and Madmen*, (London: Athlone, 1981). pp. 109-115.

[40] Edwin Chadwick, *The Evils of Disunity in Central and Local Administration: Especially with Relation to the Metropolis: and*

also on the New Centralization for the People together with Improvements in Codification and in Legislative Procedure, (London: Longmans, Green, 1885). p. 2. Chadwick cited a maxim of Jeremy Bentham: "Always to do the same thing in the same way, choosing the best, and always to call the same thing by the same name."

[41] Val Plumwood, "The Concept of a Cultural Landscape," Ethics and the Environment, Vol. 11. No. 2. 2006. pp. 115-150. Plumwood has conjectured the opportunity costs of blindness to environmental services contributed by non-human agencies under British colonial traditions.

[42] Cass R. Sunstein, *Legal Reasoning and Political Conflict,* (Oxford: Oxford University Press, 1996). pp. 35-61. See also: Anthony J. Langlois, *The Politics of Justice and Human Rights: Southeast Asia and Universalist Theory,* (Cambridge: Cambridge University Press, 2001). pp. 100-124. Fred D'Agostino, Personal Correspondence, (14 January, 2008). D'Agostino proposes a similar idea which he calls "shallow consensus," manifest in deliberate vagueness as a basis for cosmopolitanism.

[43] Thomas Homer-Dixon, The Upside of Down (Catastrophe, Creativity, and the Renewal of Civilisation), (Melbourne: Text publishing, 2006). p. 223.

CHAPTER THIRTEEN

Indigenous People's Hunting Issues and Environmental Ethics: A Contextual Observation in Taiwan

Yih-Ren Lin

Introduction

Discussion about hunting by indigenous people has long been an issue of social concern in Taiwan. In this discussion, indigenous people's hunting culture and animal liberation are particularly controversial issues.[1] From the 1990s until the present, at least nineteen PhD dissertations and master's theses have focused on indigenous people's hunting issues. Though this may not seem like a large number, it is evident that indigenous people's hunting has continually been a matter of social concern. The recent Maqaw National Park controversy[2] and the wind-fallen beech tree event in Smangus[3] point to the conflict boiling on the surface between government's natural resources management effort and indigenous rights. However, considering the deeper layers of the cultural matrix, the true crux of the issue may be the social legitimacy of indigenous ecological knowledge. As part of this, hunting is undeniably a symbol of the struggle of indigenous groups to attain rights.[4] On the other hand, out of respect for wild animals, conservationists and animal liberation groups aggressively oppose indigenous people's hunting activities on the basis that modernization has caused indigenous societies to lose the traditional knowledge they once held. Are the principles held by indigenous activists who advocate cultural hunting rights and those of the conservation and animal liberation movements fundamentally different? What kinds of actual circumstances have affected the development of the indigenous people's hunting rights, conservation and animal liberation movements in Taiwan? In the midst of the realities of social controversy, the goal of this study is to find a foundation for rational dialogue based on ethical demands.

Late in his life, the American forest service official, accomplished hunter and wildlife scholar, Aldo Leopold (1949), summed up his lifetime of experiences and interactions in the wilderness from the point of view of environmental ethics. The book *Sand County Almanac* gave a vivid and in-depth account of an ethical philosophy of the relationship between man and nature. Leopold emphasized that all creatures exist in an interdependent state of ecological holism. J. B. Callicott was a meticulous student of Leopold's land ethic philosophy, as well as an influential American environmental scholar. Since his style of writing continually engaged practical issues, his work often became the center of academic debate. Since 1980s, a paper by Callicott (1980) in the journal *Environmental Ethics* sparked a series of discussions about environmental ethics and the animal liberation movement, the content of which is closely related to the issues of indigenous people's hunting and conservation. It is believed that this series of academic dialogues can help to resolve the current ideological predicament of indigenous people's hunting, conservation and animal rights in Taiwan. Thus, the following section begins by further discussing this debate.

The Environmental Ethics/ Animal Liberation Debate Sparked by J. B. Callicott

In 1980, J.B. Callicott published an article titled "Animal Liberation: a Triangular Affair," in the well-known environmental philosophy journal, *Environmental Ethics*. In the article, Callicott points out that according to traditional moralist philosophy, only humans possess reason, interests, self-cognition, language ability and an imagination of the future. Only because of these attributes do humans have the capacity to obtain and administer rights. However, in recent times, this sort of understanding has been challenged by animal liberationists, who point out that not all humans possess the above mentioned abilities, e.g. the mentally disabled, and humans in a long-term comatose state. Referred to as ethical patients, traditional moral philosophers have bestowed rights upon such individuals nonetheless. In contrast, more and more biological research suggests that some non-human animals possess high levels of intelligence and cognitive ability, e.g., dolphins and chimpanzee. As a result, animal liberationists argue that traditional philosophy is based on a sort of anthropocentric speciesism. They also hold that this kind of viewpoint does not stand up to philosophical critique. Naturally, the discourses and actions of animal liberationists also arise in the discussion of ethics. The positions of some animal liberationists are based on utilitarianism. For example, philosopher Peter Singer has pointed out that the ability to think morally is an important prerequisite of cognitive ability. Though animals lack this kind of ability, they do suffer. Thus, they should fall within the range of ethical considerations. According to the utilitarian position, suffering is a kind of crime, and the high moral position should therefore pursue the happiness of the majority. In other words, such a stance should attempt to prevent suffering. Therefore, the utilitarian believes that philosophy should be extended to include non-human animals.

Whether their positions are based on animal rights or utilitarianism, animal liberationists' perceptions of the status of animals vary from those of traditional moralists. This has led to a number of debates (see Singer 1977, Regan 1983, Taylor 1986, Callicott 1980, 1988, 1989, 1998, Jamieson 1998). It is important to note that Callicott believes that there should be a third alternative added to the debate between traditional moral philosophy and animal liberation. That third alternative is the land ethic proposed by Leopold. However, Callicott makes it clear that the land ethic should not be confused with the animal rights and utilitarian positions of animal liberationists. This could lead to the misunderstanding that animal liberation is an environmental ethic. Callicott gives a noteworthy description of how to use the land ethic to modify traditional moralist philosophy and the animal liberation movement. However, his work emphasizes the differences between the land ethic and animal liberation, and eventually, strays away from the former issue, resulting in an accidental change of focus. The discussion that follows emphasizes the former efforts by Callicott. However, in the interest of completeness, it is necessary to touch on the differences between the land ethic and animal liberation that Callicott described. These can be summed up by two main differences: the difference between community and individual values and the varying interpretations of suffering.

First off, Callicott points out that Leopold's land ethic emphasizes the values of ecological holism. This viewpoint is reflected by the following oft-quoted line:

> A thing is right when it tends to preserve the integrity, stability, and beauty of the <u>biotic community</u>. It is wrong when it tends otherwise. (Leopold 1949)

Though humanitarians and radical moralists disagree about whether or not non-human animals possess moral rights and moral standing, Callicott believes that both sides are still similar in that they discuss the rights of individuals. Callicott describes this as an atomistic or distributive theory of moral values. He opposes such a theory on the grounds that it runs counter to the holistic or collective ecological ethic proposed by Leopold. At the same time, Callicott points out that the tension between traditional moralists and animal liberationists can be ameliorated by the third alternative offered by Leopold's land ethic. This is the primary meaning of the term "Triangular Affair" in the title of Callicott's article.

What kind of resolution does Callicott suggest? He points out that traditional humanitarianism states that humans have higher status than other animals, and only humans have moral rights. Thus, based on their own morality, humans possessing moral rights should work in the direction of treating animals in a humane manner. However, humanitarians insist it is not proper to bestow moral rights upon animals. In regard to this point, Callicott believes that the land ethic takes a different position on the relationship between humans and non-human animals within an ecosystem. For example, highly intelligent animals who can perceive suffering like dolphins and chimpanzee have the same rights. As for the difference between humans and non-human animals, Callicott points out that the land ethic takes the same position as traditional moralism. However, it is important to note that the reason the land ethic takes this stance is very different!

In contrast, radical animal liberationists suggest that humans should stop thinking of themselves as superior to other animals. They believe that this is a speciesist point of view and thus promote either the moral rights of animals or utilitarianism. From the point of view of the land ethic, humans should not consider themselves as the masters of nature. Due to the interdependence of all species, it is important to protect healthy relationships within ecosystems. Looking at the position of humans in nature, the land ethic and animal liberation deliver similar critiques of human chauvinism. However, the motives and theoretical positions of the two stances differ greatly. Based on this analytical framework, Callicott illustrates the similarities and differences between the land ethic, ethical humanism and humane moralism advocated by animal liberationists. In particular, he points out that atomistic and distributive discussions of individuals and rights inevitably restrict moral issues to competitive views of individual profit and rights. Thus, such discussions are constantly trapped within a state of contradiction. In the conclusion to his study, Callicott sums up the aforementioned differences as follows:

> Humane moralism remains firmly within this modern convention and centers its attention on the competing criteria for moral standing and rights holding, while environmental ethic locates ultimate value in the biotic community and assigns differential moral value to the constitutive individuals relatively to that standard. (Callicott, 1980)

In the 1980 article, Callicott also discusses interpretations of suffering. He does not completely agree with the singular goal of animal liberationists to reduce the level of suffering of all beings that are able to perceive suffering. In an additional article, "Animal Liberation and *Environmental* Ethics: Back Together Again," Callicott (1989: 33-34) gives an even clearer explanation of how the land ethic differs on this point. He states:

> To live is to be anxious about life, to feel pain and pleasure in a fitting mixture, and sooner or later to die. That is the way the system works. If nature as a whole is good, then pain and death are also good. Environmental ethic in general require people to play fair in the natural system…Instead of imposing artificial legalities, rights, and so on nature, we might take the opposite course and accept and affirm natural biological laws, principles, and limitations in the human personal and social spheres…Such appears to have been the posture toward life of tribal peoples in the past. The case was relished with its dangers, rigors, and hardships as well as its rewards; animal flesh was respectfully consumed; a tolerance for pain was cultivated… It is impossible today to return to the symbiotic relationship of Stone age man to the natural environment, but the ethos of this by far the longest era of human existence could be abstracted and integrated with a future human culture seeking a viable and beneficial relationship with nature. (Callicott, 1989)

In the explanation given above, Callicott rejects the utilitarian approach of pursuing the happiness of all humans, or possibly non-humans as well, and minimizing their level of suffering. Importantly, this does not mean that humans can arbitrarily cause the suffering of other species. Rather, he emphasizes that when pursuing the health and stability of the individual, we should consider the ecological whole. From this point of view, he believes that under certain circumstances, suffering is not necessarily entirely evil. Just as when a wolf hunts its prey, the prey is forced to cope with brief and intense suffering. However, the harvest of the hunt can nourish the wolf's entire body, allowing it to continue living. The holistic ecological position that "everything is connected with everything else" is the foundation of Callicott's criticism of the animal liberationist stance. At first glance, this position seems very "inhumane" and "disrespectful of life." For instance, it acquiesces to the existence of hunting. It is important to note that the kind of hunting that is acceptable is carried out within the bounds of the natural order. In other words, Callicott does not believe that hunting is tantamount to the reckless infliction of suffering. Callicott's idea of the natural order includes the entirety of human society. He mentions the ecological knowledge of the indigenous people of North America (Callicott 1989), pointing out that indigenous people's hunting has its own standards. That is, the indigenous people do not hunt animals unconditionally. More importantly, to discuss hunting in a meaningful way, it is necessary to explore the social and cultural context within which it takes place. Considering the interdependent and intrusive relationship between the natural environment and human society, Callicott goes on to support the concept of animal liberation proposed by another moralist scholar, Mary Midgley (1983), who used the term, "mixed community."

Callicott (1989) points out that the insight offered by Midgley not only agrees with the holistic ecological position that "everything is connected with everything else" within the context of biological evolutionary relationships, but also overcomes the dichotomy between nature and man (or human society). The concept of "mixed communities" does not only offer insight into the animal liberationist view of pets (also referred to as companion animals) and domestic animals, but can also provide a critique of the current discussion of the economic animal. After all, granted that humans and other animals exist socially, there should be a serious discussion of the social and cultural modes of interactions between humans and animals, as well as how such modes change. Such a discussion should not merely consider the moral status and rights of animals alone! Returning to the interpretation of suffering, to understand both suffering and happiness, it is necessary to take natural and social contexts into account. In other words, when it comes to suffering and happiness, only by incorporating these contexts can we make moral judgments about what should and should not be done.

On the surface it seems that Callicott wisely draws Leopold's land ethic into the tension between traditional moralists and animal liberationists. Moreover, using the land ethic as a third alternative, Callicott resolves the ambiguity between environmental ethics and animal liberation. In reality, this is not the case! The article by Callicott quickly sparked a debate that spanned nearly twenty years (Warren 1983, Regan 1983, Midgley 1983, Sagoff 1984, Callicott 1989, Jamieson 1998, Callicott 1998). Because of this unexpected debate, in the follow up article, "Animal Liberation and Environmental Ethics: Back Together Again," (Callicott 1989:49) Callicott admits feeling slightly regretful shortly after writing "The Triangular Affair." He states:

> From a practical point of view, it would be far wiser to make common cause against a common enemy—the destructive forces at work ravaging the non-human world—than to continue squabbling among ourselves.

Callicott admitted that his overemphasis of the differences in "The Triangular Affair" brought about negative results. Consequently, new rifts emerged between allied social movements and the urgency of working together to face destructive forces was neglected. Due to the limitations of this study, the following discussion focuses on the recent follow up articles and reviews of Callicott's "The Triangular Affair" by Dale Jamieson. This is because Jamieson's critique sums up the issues of the debate mentioned above. Moreover, his position is relevant to the animal rights/indigenous people's hunting controversy discussed in this study.

First, Jamieson explores the similarities between animal liberation and environmental ethics within the context of social movements. His main point is that both of these social movements emerged after WWII in the 1960s as part of a wave of thinking opposed to mainstream Western culture. Both movements objected to "sacrificing everything else to build weapons." They also both criticized the tendency of material culture to "view humans and animals as replaceable commodities," and even more seriously, to believe that "technology can solve all problems." As illustrated by these issues, Jamieson believes that environmental ethics and animal liberation share common points. This differs from Callicott's proposition that the two are based on completely different concepts. Jamieson rightly points out that the dispute between environmental ethics and animal liberation should not merely be a matter of philosophical analysis. Within the context of social movements, Jamieson shows that the similarities between the two movements outweigh the differences. Thus, he believes that Callicott exaggerated their differences.

Second, Jamieson also points out that there is no one definitive school of environmental ethics. Moreover, animal liberation is also based on a wide array of theoretical positions. Jamieson gives a brief review of the historical development of environmental ethics. He touches on important discussions within the field, such as Richard Routley's (1973) point that nonsentient entities have their own value distinct from that of humans. On this subject, Tom Regan (1981) suggests that environmental ethics should be clear about the moral standing of these nonsentient entities. In his earlier work, John Passmore (1974) disagrees with both of these positions. In another important discussion, the concept of "deep ecology" promoted by Norwegian scholar Arne Naess (1973) emphasizes the "rights of nature," making yet another unique contribution to the field of environmental ethics.[5] Jamieson shows that the land ethic promoted by Callicott in the 1980s was just one of many schools of thought. In fact, among the group of scholars working in close proximity to each other, one of Callicott's friends, Holmes Rolston III, was also an adherent to the principle of ecological holism. At different times, both Callicott and Rolston served as the President of the International Society for Environmental Ethics. However, the two had very different opinions about whether or not nature has inherent rights.

Just as environmental ethics is made up of a variety of positions, the animal liberation camp is also based on a diverse array of theoretical stances. Classic works such as *Animal Liberation* by Peter Singer, *Animals and Their Moral Standing* by Stephen Clark, *Animal Rights and Human Morality* by Bernard

Rollin, *Animals, Men and Morals* edited by Godlovitch, *The Case of Animal Rights* by Tom Regan and *Animals and Why They Matter* by Mary Midgley discuss the various discourses of animal liberation, including Utilitarianism, Kantianism, free will and Communitarianism. Thus, animal liberation is the amalgam of a diverse spectrum of thought, rather than a single set of values and beliefs. Jamieson optimistically believed that the various positions of animal liberation could easily find a way to engage in dialogue with environmental ethics. Furthermore, the history of the social movements provides actual examples of the two sides cooperating on certain issues, such as in the opposition to dumping of toxic waste in rivers, air pollution, the threat of the extinction of certain species caused by whale hunting, the clearing of the rain forest to produce beef, etc. The environmental ethics and animal liberation camps actually worked together on these issues. In the article written in 1980, Callicott clearly neglected these cases and placed too much emphasis on the land ethic.

In addition to the social movement context, Jamieson looked for points at which the philosophies of the two sides intersected. One important point is the human versus nonhuman debate. Overall, by exploring the important social contexts in which the ideologies developed, Jamieson tried to settle the differences between environmental ethics and animal liberation. When looking at the historical development of discourses, or the diverse spectrum of ideologies, or the extension to the philosophy of value, Jamieson emphasized that the analysis should seek a "contextual understanding." Within this kind of understanding, he tried to find potential cases for real cooperation. He used the animal liberationist Gary Varner as an example. Considering the overall benefit of target species and the integrity of ecosystems, Varner suggested that so-called "therapeutic hunting" could endanger individuals within entire groups. In reality, such considerations are only useful when they take into account specific social contexts, such as in the complicated case of the spotted owl (Yaffee 1994). Both environmental ethics and animal liberation are comprised of a wide spectrum of complicated ideologies. Neither side advocates the dichotomization of issues. In the article "'Back Together Again' Again" Callicott (1998) praises efforts to find harmony between the two sides. He goes on to say that he has already discussed some of the points made by Jamieson in the article "Back Together Again (1989)," for instance, Jamieson's discussion of "inherent value." This being the case, Callicott states that the difference between his work and Jamieson's was that Jamieson's work was based on classical moralist philosophy and the universalization of egoism. That is, due to the principle of objectivity, people must take their own welfare into account and extend this concern to other humans (or nonhumans). In contrast, Callicott adopted the communitarian stance of Leopold's land ethic.

In sum, throughout the debate about the differences between environmental ethics and animal liberation sparked by the "Triangular Affair," there have been some hard feelings at times; however, the debate has also led to a great deal of positive outcomes. First, as a result of the bold provocations offered by Callicott, we have come to a better understanding of the ideological spectrums that make up the environmental ethics and animal liberation movements. Secondly, looking at the actual accomplishments of the social movements, both sides have gradually realized that while differences in ideological approaches may lead to conflict in some cases, there are many cases in which the positions of the two sides can work together to face even greater challenges, such as opposing the large-scale development of rainforests, particularly the development of capital intensive ranches. Third, since the concept of animal liberation that Callicott cultivated from Leopold's land ethic does not focus on the moral rights of animals, his attitude toward animals may be more acceptable to a broader spectrum of society. This is also true of non-animal and abiotic entities. This study attempts to use this kind of concept to ameliorate the tension between indigenous people's hunting, conservation and animal liberation in Taiwan.

The Triangular Affair of Taiwanese Indigenous People's Hunting, Nature Conservation and Animal Liberation

In Taiwan, the discussion of the appropriateness of indigenous people's hunting began in the mid 1980s. The discussion was started by conservationists, not animal liberation groups. At the time, there was a great deal of progress in regard to the protection of wildlife in the areas of ecological research and in the establishment of different types of nature preserves and laws intended to protect wildlife (Patel and Lin 1989). As a result, the voices calling a halt to the continuous harm brought to wild animals gained significant social legitimacy. Scholars and conservationists generally lumped hunting together with mining, the building of roads and the destruction of habitats as the main factors leading to decreasing wildlife populations. Moreover, indigenous people were placed together with property owners developing mountain regions as the joint culprits responsible for sharp drops in the populations of various types of wild animals (Wang 1986, 1987, 1988; Rabinowitz and Lee 1990; Nowell 1991).

Due to limitations in research methods during this period, academic research, the enactment of laws, governmental enforcement and public opinion all led to the stereotype that indigenous people were "wildlife killers." The majority of academic studies carried out during this period made rough estimates of the amounts of wild animals "removed" (primarily large and mid-sized mammals like mountain boar, goat, muntjac, sambar deer, civet, etc.) from the wild. These estimates were based on oral interviews with people living in mountain regions (indigenous people, forest rangers, miners, etc.) and the amounts of goods sold by stores dealing in mountain products. However, these studies failed to investigate the changing numbers of various types of wild animals still living in mountain areas.[6] In addition, these studies lacked solid evidence detailing the actual significance of indigenous people's hunting compared with other factors, such as road building, mining and the destruction of habitats, in the decrease of wildlife populations. Thus, the inference that indigenous people were the culprits responsible for decreasing wildlife populations was not so much a serious academic conclusion as it was the result of the intense sense of urgency held by wildlife conservationists! With the rising tide of wildlife conservation in the 1980s and the subsequent attempts to explain why wildlife populations were decreasing, the indigenous people could not avoid becoming scapegoats. Even more seriously, until now, there is still no conclusive evidence describing the actual impact of indigenous people's hunting on the various types of wild animals in Taiwan.

In the early 1990s, there was a slight change in the kind of discourse described above. First, the leaders of the indigenous movement began to use the "spirit of the hunter" as an important symbol and point for discussion. This kind of discourse had the effect of "decriminalizing" indigenous people's hunting after it had been subjected to a long-term period of stigmatization by conservationists. On the other hand, this discourse also implied that the protests of indigenous people were moving away from urban areas and back to the "homeland." These factors can all be viewed as the manifestations of the spirit of tribalism (Wang 2001). In 1996, Pei Jia-chyi, a professor at the forestry department at National Pingtung Institute of Science and Technology, and Taiban Sasala, a member of the Rukai people, held a conference in Wutai Township in Pingtung County. The conference was titled, "Indigenous People are the Guardian of the Mountains." First, this conference changed the role played by indigenous people in nature conservation. The negative impressions of the past were gradually transformed into an active role for indigenous people. Rukai hunters, ecologists and conservationists gathered together to form the "Rukai Eco-Conservation Foundation," aligning the "spirit of the hunter" discourse with nature conservation, which converged to create the even more meaningful discourse of the "guardian of the mountains." These discourses stressed that the longstanding and intimate connection between indigenous people and nature gave birth to a wealth of practical knowledge made up of ceremonies, taboos, stories and myths. The ecological principle that everything is interconnected is reflected in this kind of traditional knowledge. Previous discourses describing indigenous people as "wildlife killers" de-emphasized the "local knowledge" of indigenous people and seriously neglected the fact that, due to their practical and ideological contributions, indigenous people are the ultimate participants in the conservationist movement. In the 1990s, this kind of discourse did not just appear in Wutai in Pingtung County, but also began to foment in indigenous villages throughout Taiwan. The "Danayigu Ecological Park" managed by the Tsou Tribe's Shanmei village near Ali Mountain is another famous example (Chen 2002). The "Multiple Faces of Indigenous people's Hunting Issues Conference" held in Wutai Township in August 2000, not only emphasized that indigenous people are the guardian of the mountains, but also propounded the willingness and desire to actually manage mountain resources (Lin 1992; Fu 1997; Huang 2001; Lin, 2001; Taiban Sasala 2002). These actions directly clashed with Taiwanese society's impressions of indigenous people's hunting and challenged the regulations set up to conserve mountain resources.

The 1990s was a period of continual political transformation in Taiwan. During this period, the issue of indigenous people's hunting went through changes as well. On the one hand, as mentioned above, indigenous activists attempted to alter the "wildlife killer" stereotype held by the conservationist movement, and also sought to find a cultural foundation for the indigenous movement based on the "spirit of the hunter" discourse. On the other hand, other social movements also began to arise in Taiwan in the 1990s, one of which was the animal liberation movement discussed in this study (see footnote 1). In 1992, the popularity of a cruel and inhumane game involving the "fish hooking" came to the attention of Master Chao-hui, a member of the Chinese Buddhist Association. She made efforts to gather leaders of different religions and other social figures together to petition the Premier, Hau Po-tsun. Ultimately, an executive order was issued forbidding the practice (Lin 1998: 90-130). These events eventually led to the formation of the Life Conservationist Association. In an article titled, "Buddhists on Earth Appear and Deliver the Teachings: From Animal Liberation to Buddhist Feminism," Master Chao-hui (2002) recalled:

> After assessing the circumstances, I believed that I was not strong enough on my own. It was necessary to have a longstanding association to act on behalf of animals. Such an association can negotiate to obtain legal protection and security for animals at a systematic level. Moreover, it can stop members of society from torturing

animals. On another level, through publications, propaganda and education, the association must work to spread the principle that "all beings are equal" to society at large. These efforts can correct the chauvinistic tendencies of humanity.

Master Chao-hui expressed a deep concern for animals. Through the formation of an organization, legislative action, propaganda and education, she hoped to rectify the lack of concern for the welfare of animals commonly held by society. At the time, the association's secretary-general, Master Wu-hong, along with members of the alliance's secretariat and other volunteers began many movements and campaigns that would later become pivotal, such as disseminating the association's newsletter and the "Voice of Taiwan's Animals"; protesting the production and sale of bear gall bladders in mainland China; participating in the amendment of the Wildlife Conservation Law; promoting the "don't eat, don't raise and sell wild animals" campaign; bringing together various environmental organizations to form the "Eco-Conservation Alliance"; working to save wetlands; opposing the legalization of race horse gambling; translating the classic animal rights book *Animal Liberation*; forcing officials to deal with the stray dog problem; and participating in the drafting of animal welfare laws, etc. Through these actions, it can be said that the members of the secretariat made full use of the surge of social progress that occurred after the lifting of martial law. The association's overall results were gained through appropriate strategy and the reemergence of the media. Generally speaking, a new platform had emerged to help animals in Taiwan (Lin 1999). However, it should be noted that these seemingly complicated actions were the products of an important central principle, that being the Buddhist principle of "protecting life." This is precisely why the movement to protect animals in Taiwan should not be confused with the philosophy and actions of the animal rights movement in the West. Though the book *Animal Liberation* by Peter Singer has had a certain degree of influence in Taiwan, and Peter Singer himself has visited Taiwan, the animal protection movement in Taiwan must be viewed within the actual social context of the times. Master Chao-hui (2002) has expressed this point clearly:

People often use the term "animal rights." Actually, the idea of "animal rights" comes from the West. In Buddhism, the concept of "protecting life" has its own complete theoretical structure. However, the reason for "protecting life" is not derived from the idea of "animal rights." "Animal rights" is based on the concept of "inherent" rights, which is very vague and difficult to justify. Legally speaking, in practice, "rights" are often accompanied by the concept of "obligation." Thus, if they are not careful, proponents of "animal rights" fall into the trap of the argument stating that, "animals do not qualify to obtain the rights enjoyed by humans."

In the early 1990s, Master Chao-hui and the other members of the Life Conservation Association began to work toward animal liberation based on the Buddhist practice of "protecting life." The association participated in a diverse range of activities that included elements of Western-style animal rights and environmentalism. For example, they showed concern for stray dogs and worked to save wetlands on the west coast of Taiwan. To the surprise of many observers, various types of social movements that may appear very different from the Western perspective were merged together. Without looking at the historical development of the LCA and the changing social circumstances in Taiwan, it would certainly be difficult to classify these various actions by using the standards of the Western animal liberation movement. This might be why Master Chao-hui offered the above critique of "animal rights." Even more importantly though, the above statement represents the connection between the discourse of Master Chao-hui and the LCA with Buddhist thinking, as well as the active attempt to create a dialogue with animal liberationist philosophy from a Buddhist perspective. These sort of attitudes emerged within Taiwanese Buddhism as part of Taiwanese society in the 1990s, with its subjectivity emerging through the integration of different philosophies combined with active participation in social issues. The same kind of examples can be derived from Master Yin Shun's idea of "the Buddha on Earth" as described by National Taiwan University professor Yang Hui-nan in an article in the magazine *Con-temporary Monthly*, and the debate that followed (Yang 1994, Lu 1995, Lin, 1996; Lin 1999). Whether the argument is proper or not, the following words by Master Chao-hui clearly indicate her aggressive attitude to engage with the social issues from a Buddhist point of view:

> In the book *Animal Liberation*, Peter Singer argues meticulously, promoting the theory that people should expand morality to include animals. He believes animals should have rights and rejects any reasons for slaughtering animals (especially ideas like, "people have reason and intelligence and animals do not" and so on). It is not necessary for animals to "fulfill obligations" in exchange for the right to live. Just understanding that animals, like humans, feel the pain of slaughter and torture, is reasonable enough to treat animals equally. His position has been classified as "teleological" or "utilitarian." We believe that even if his original argument resembles secular teleology, however, "utilitarianism" is the closest thing to the spirit of compassion or "protecting life" within Buddhism. This is because his focus is to "minimize the suffering of animals," rather than "maximize the happiness of humans." In this regard, he has transcended the "anthropocentric" way of thinking and reached the similar conclusion to the Buddhist way of thinking, that "all beings are equal."

In sum, when looking at the issues of indigenous people's hunting and animal liberation, it is necessary to look beneath the surface and closely analyze the various positions of the activists involved. Even more importantly, we must understand the relationship between the discourses of these activists and the changes within Taiwanese society. From the above analysis, the rough outlines of three subjects emerge: conservationists, indigenous people, and Buddhist organizations (this study focuses primarily on the discourse of the Life Conservation Association). As Taiwanese society went through a series of tumultuous changes, all three groups attempted to make their voices heard.

Moreover, as these groups have developed into social subjects, traces of their actions can be detected in the tension between indigenous people's hunting and animal liberation.

Are the Issues of Indigenous People's Hunting, Nature Conservation and Animal Liberation Truly Antagonistic?

What exactly is the connection between the indigenous people's hunting, conservation and animal liberation movements in Taiwan? To answer this question, it is necessary to return to the involvement of the Life Conservation Association in the amendment of the Wildlife Conservation Law in 1993. The amendments made in 1993 were certainly influenced by the commotion caused in Taiwanese society by the rhinoceros horn event.[7] In the amendment process, the LCA had two main points of dispute. First, from the viewpoint of animal liberation, they opposed proposals to allow hunting. Second, based on the idea that wild animals should live in nature, they were opposed to allowing people to raise wild animals. Thus, the LCA brought together dozens of environmental groups to establish the "Eco-Conservation Alliance." Bringing forward the so-called people's version of the Wildlife Conservation Law, the alliance actively intervened in the amendment process. They took a very stern position on the practice of raising wild animals. To achieve their goal, the LCA demanded that government officials create a sunset clause to end the practice of raising animals in certain period. During the amendment process, the "Eco-Conservation Alliance" won crucial victories on these two fronts. First, in regard to indigenous people's hunting, the result can be seen in the 21st Article of the Wildlife Conservation Law. The amendment stipulating the relationship between wildlife and indigenous people's hunting is listed in clause five. It states, "Taiwanese indigenous people living in lands reserved for their use who find it necessary to hunt, slaughter or exploit wild animals due to their traditional cultural festivals," must apply for permission from related administrative units. Basically, this legal amendment, which was based on a static understanding of indigenous people's hunting culture, was made in order to prohibit any commercial activity related to indigenous people's hunting, such as stores selling products from mountain regions, under the pretext of respecting "traditional culture." However, the amendment neglected the fact that in the past, hunting was a part of indigenous people's traditional way of life. It was not merely ceremonial. More importantly, hunting was a way to meet basic material needs, such as acquiring animal proteins, skins, bones, etc.

In addition, the prohibition of the raising of wild animals is included in the 31st Article, which stipulates, "Prior to NPA announcement, all persons engaged in breeding protected wildlife or exotic wildlife dangerous to the environment, or those possessing protected wildlife products as determined by the NPA's announcements designating protected wildlife shall fill out a data card and keep records with their municipal or county authorities and update these records within a certain time period after any change in status." The article goes on to state, "Those who have registered according to the requirements may be allowed to continue to raise or hold their wildlife or wildlife products, but no breeding shall be allowed except for academic research or educational purposes and with the approval of the authorities." The amended article also contains a so-called sunset clause which demands that, "Those who have engaged in raising or breeding any of the wildlife listed in Paragraph 1 before promulgation of these amendments shall be assisted by the authorities to cease raising or breeding the animals and change occupations within three years of the date of promulgation. If necessary, the animals may be purchased by the authorities. The purchase of the animals shall be done in an appropriate and safe manner and any wildlife may be sent to domestic or foreign educational or academic institutes or zoos, or taken in and cared for by an organization considered appropriate by the authorities and commissioned by them." In the amendment process, the "Eco-Conservation Alliance" organized by the LCA effectively achieved their goal of prohibiting the exploitation of wild animals. Interestingly, this result had the effect of softening the conservative views of some conservationists toward the "use" of nature. The concept generally referred to as "preservation," aims to protect primitive natural areas that are untouched by humans, whereas, the successes of the still growing LCA organization were based on the concept of "animal liberation." As expressed by the LCA's founder, Master Chao-hui, the organization cannot tolerate the suffering of any beings and stands in opposition to human chauvinism. This stance is heavily influenced by the Buddhist doctrine of "protecting life." In a subsequent article, Master Chao-hui (2001) went on to describe this position even more clearly:

> From a Buddhist perspective, the collective experience of disasters cannot be understood by discussing the "moral indeterminancy" of animals. This is because emotionless entities do not leave mental karmic traces. However, to experience first hand the law of dependent arising and the interdependent nature of existence and to have a feeling of gratitude, to love and protect the natural environment which "sustains mammals and all living beings," we cannot neglect their livelihood merely on the grounds that they "do not fulfill an obligation." We cannot have doubts based on the concept of "obligation," but should work toward our goal indirectly from the point of view of "protecting life."

In the above statement, Chao-hui clearly expresses that according to the Buddhist doctrine of dependent arising, abiotic entities, like minerals and flowing water, which are emotionless entities outside the bounds of the karmic cycle of life, have an interdependent relationship with living "emotional beings," and should thus be cherished. Moreover, the protection of living beings should not be based merely on the relationship between one being and another. By abandoning the traditional moralist position of "rights and obligation," Chao-hui demonstrates that the Buddhist principle of "protecting life" is based on the interdependence of all beings. Importantly, her stance is derived from the Buddhist belief in "dependent arising." What exactly is dependent arising? She explains as follows (Chao-Hui 2001):

> How is it that people appear to take the form of individual entities but are not "disconnected" entities?

> From the point view of dependent arising, no phenomena (including the sentient) are isolated, existing independently. There must first be the "interaction of dependent beings" before phenomena can take form. Thus, the principle by which one entity arises is the same as that of other entities, in that they make up a complicated and "interdependent" network. Under this pretext, cause and effect successfully supports the connections between one entity and another. How then would it be possible for a subtle and unimpeded flow not to exist? Therefore, respect for other life is not simply some naive emotional state. This kind of respect is a law causing the natural flow of emotion.

The Buddhist concept of dependent arising was originally used to explain the reason for the origins of all beings. Here, Chao-hui extends the argument to the social issue of animal liberation. Chao-hui's interpretation of the law of dependent arising is intimately connected with the trajectory of her social movement. In all efforts of the LCA, Chao-hui is the guiding force behind the organization's discourse. Furthermore, she actively integrates Buddhist teachings with her positions on social issues. The following quotation reflects this kind of attitude even more clearly:

> The ecological idea of the "diversity of life" coincides with Buddhism. More specifically, this refers to "the diversity and uniqueness of life and the complexity of species in ecosystems." Only those with a deep understanding of dependent arising can glimpse the reality that "all things arise interdependently and are inherently empty" and transcend the self to see the interdependent nature of the environment. From this understanding, they respect the value of the existence of all forms of life and look upon them equally. In this way, people can develop selflessness, reciprocity, gratitude and modesty and take compassion as their philosophy in life.

Chao-hui attempts to integrate the concepts of ecological diversity, dependent arising, protecting life and the equality of all beings in her stance on animal liberation. This kind of attempt also explains the idea of animal liberation in the social context of Taiwan. The situation in Taiwan cannot be understood simply through the discourse of the animal liberation movement in the West. The social context of Taiwan has certainly produced its own unique characteristics, particularly in regard to the interpretation of Buddhist teachings. Returning to the process of amending the Wildlife Conservation Law, on the surface, it appears that the two main positions of the LCA stand in opposition to different groups, one being indigenous people and the other being people involved in raising and breeding wild animals. The actual motive for opposing these two groups is the same, that being, to oppose the exploitation of wild animals by humans, and to promote the idea that there is good reason for wild animals to coexist in nature. This is the position which Chao-hui refers to as "protecting life." The nature conservationists discussed above also oppose the exploitation of animals by humans in their efforts to prohibit hunting. However, the motives that lie behind this position are very different.

The animal liberation stance expounded by Chao-hui through the social actions of the LCA gradually came into conflict with the "wise use" trend arising within conservationist movement in Taiwan in the 1990s. The concept of "wise use" differs from the idea of wildlife "preservation" described above when it comes to the prohibition of hunting. The idea of wise use emphasizes the balance of ecosystems in the management of natural resources. In the wave of support for wise use, the staunchest advocate has been National Pingtung University of Science and Technology ecologist, Pei Jia-chyi. He supports allowing indigenous people's hunting on a "conditional" basis, and he is also a supporter of the "guardian of the mountains" discourse. However, from the point of view of the indigenous movement to attain hunting rights, it seems that one problem has been followed by another. As the original discourse of "wildlife killers" has been replaced by "guardian of the mountains" within the conservation movement, the "protecting life" position of the animal liberation movement toward wildlife has emerged as a dark house in the process of amending the Wildlife Conservation Law. Moreover, the "protecting life" camp allied itself with conservationists in support of "preservation." This has led to tension between the beliefs and positions of the indigenous movement and the Eco-Conservation Alliance. This kind of tension can be detected in the following words by Taiban Sasala (2000):

> In Taiwan, indigenous people's hunting culture is like a riddle that bewilders academia and the proponents of conservation. It even puzzles some of the young indigenous intellectuals who are involved in the indigenous rights movement. Is hunting really only related to the problem of conservation and the guaranteeing of animal rights? Or is there another possibility? As conservation is becoming a mainstream tide of thought in the modern world, haven't people ever considered that hunting can become an effective force to help in conservation? This dichotomy in the thinking by most people has already created a gap between conservation and hunting which is very difficult to cross. Conservation is conservation and hunting is hunting. The two can never be compatible.

Taiban Sasala is clearly aware of the tense relationship between indigenous people's hunting and nature conservation. However, this sort of anxiety has a positive side! It is an effort to escape the opposition between the two sides. Basically, ecologists' concerns with indigenous people's hunting lie in the sustainability of various wildlife species. There are varying opinions within the field, such as the concepts of "preservation" and "wise use" mentioned above. These two camps have different attitudes toward indigenous people's hunting. The former is inclined to prohibit hunting while the latter advocates allowing hunting on a conditional basis. This latter view is expressed in Taiban Sasala's suggestion that "hunting can become an effective force to help in conservation." The question is how is this possible? Taiban Sasala (2002) points to the "Experimental Plan to Use Hunting to Manage

Natural Recourses" conference held in Wutai Township in Pingtung County as a possible direction to follow:

> The hunted boars can be gathered and managed by the township office. When tribe members have weddings or other celebrations they can apply to use the boars free of charge. The goal is to use seasonal hunting activities to estimate the population of boars in the given region. This can be used as a reference to determine the feasibility of operating hunting business in the future. The Rukai believe this is a self devised kind of "new economy." On the one hand, this kind of self-help economic model can bring hunting activities within the reigns of formal management and help discourage some of the illegal activities currently being carried out in mountain regions. On the other hand, with this kind of method, hunting activities can avoid being drawn into the commercialized logic of non-mountain regions. Control of the market can be placed in the hands of indigenous people. Simultaneously, we hope to use modern concepts of management to protect the environments of mountain regions and create a miracle of cultural regeneration and tribal development for the Rukai people.

In regard to this issue, the chief editor of *Newton Magazine*, Huang Yi (2001), who had previously generated substantial public opinion about this issue, maintained a skeptical attitude. She said:

> Indigenous people emphasize that they are no longer willing to be manipulated by the Han capitalist economic system. They want to contribute a type of knowledge for development based on coexistence with the natural environment. This is great. However, if we look at the actual amount of hunted animals sold to tourists and restaurants, further relaxation of hunting regulations may only immerse indigenous people deeper in the commercial thinking of the Han, making it more and more difficult to escape the dependence and control

of the capitalist market. This is something that our indigenous compatriots should think about very carefully.

Huang Yi's statement points to a blind point of the very popular notion of indigenous ecological wisdom (Lin 2001). That is, this discourse ignores the fact that indigenous societies have been influenced by the forces of modernization, leading to a rapid pace of change. Due to the impact of mainstream social institutions, every aspect of indigenous society has changed, including language, culture, religion, politics, economics, etc. (Chen 1973, Huang 1993). Similar in part to the argument presented by Huang Yi, Chang Long-Sheng (2002), former active participant in the planning of the National Park System and member of the Council for Economic Planning and Development, points out that this can be viewed as the most common doubt voiced by the "preservation" camp:

> Poaching is still quite common in Taiwan. There are indigenous and non-indigenous poachers. The goal of poaching is not to protect cultural heritage. It is for profit. This sort of profit motive is not accepted by international society...Therefore, if a clear line is not drawn between cultural tradition and commercial activities, if there are no strict guidelines, in the future, there are those who will carry out commercial activities in the name of preserving tradition. The demographic makeup of the people entering mountain areas is complicated. However, indigenous people will bear the responsibility for injustices. Is this not something that should be considered carefully?

Actually, from Huang Yi's point of view, conservationists who adhere to the concept of preservation are not concerned with the problem of "traditional" indigenous people's hunting. Rather, they are concerned about whether or not the hunting that has already become "untraditional" exhibits ecological knowledge that corresponds with the preservation of nature. However, there is another matter worth noting. Huang Yi does not take a completely oppositional stance. She only warns not to romanticize the ecological knowledge of indigenous people and not to ignore the influence of the changes that have taken place in indigenous society. It cannot be denied that there will still be tension regardless of whether or not indigenous people's hunting is allowed. Moreover, this tension only becomes more complex when indigenous people emphasize the discourse of "environmental justice." From the point of view of the indigenous people's movement, "environmental justice" is the discourse most commonly used to counter opposition to indigenous people's hunting. The following statement by Chi Chun-chieh (1996) is very representative of this position:

> As the earliest managers of the ecological environment in Taiwan, indigenous people developed the most primitive and effective ecological ethics and preservation measures. The hunting culture which included hunting zones, hunting groups, divination, taboos, ceremonies and sharing, once allowed the indigenous people and wild animals to maintain a long-term "dynamic balance" in the mountains of Taiwan. Today, since indigenous people have become the "dusk people" of Taiwan, this ancient culture of conservation has been severely damaged. At the same time, the mountain and wilderness ethics in Taiwan have vanished. All that is left are the controversial polices enforced by the government to manage the conservation of wildlife. In this process, indigenous people have not only been sacrificed once again, they must take responsibility for a crime they did not commit.

Speaking on behalf of vulnerable ethnic groups, Chi Chun-chieh points out that in the past, advocates of nature conservation accepted the mainstream ideas about conservation from the West, ignoring the fact that different ethnic groups have varying concepts of nature. In particular, contempt for cultural ecological viewpoints expressed by indigenous people's hunting practices such as hunting zones, hunting groups, divination, taboos and ceremonies injures indigenous people on two levels. On one level, it does not recognize the legitimacy of their cultural knowledge. On another level, it stigmatizes their hunting practices. Members of the indigenous rights movement identify strongly with the idea of "environmental justice." Thus, yet another variable is added to the already tense relationship between the indigenous people's hunting and animal liberation movements. However, it is very important to note that Chi Chun-chieh does not have an overly romanticized view of the traditional ecological knowledge of indigenous people. In fact, he states that, "since indigenous people have become the "dusk people" of Taiwan, this ancient culture of conservation has been severely damaged." His main point is that due to their weak position in society, indigenous cultures, societies and economies have all been oppressed. As a result, he does not unconditionally promote the discourse of indigenous ecological knowledge!

All of the various discourses making up the "hunting" and "anti-hunting" camps take different positions. The pro-hunting camp gains support from the "indigenous rights," "wise use" and "environmental justice" discourses. In the 1980s, conservationists espousing "preservation" supported the "anti-hunting" camp. In the 1990s, this camp gained support through the concept of "animal liberation" as interpreted from a Buddhist viewpoint. On the surface, in the process of amending the Wildlife Conservation Law, these two camps appeared to stand in stark opposition to each other. However, a closer look at the statements made by supporters of both sides reveals that there is some room for dialogue. For example, the "indigenous people are the guardian of the mountains" discourse promoted by Taiban Sasala finds common ground between the indigenous people's hunting and nature conservation movements. Chi Chun-chieh's emphasis on giving a voice to vulnerable indigenous groups has given new life to ecological culture in Taiwan. Conversely, the thoughts shared by Huang Yi are not opposed to traditional indigenous culture, but rather question how changing cultures can preserve the ecological wisdom of their ancestors. Chao-hui, on the other hand, uses the argument that all beings are interconnected and promotes her view of animal liberation based on the Buddhist concept of "protecting life." In the following quotation (Lin 1998), she gives an even clearer description of "protecting life" and indigenous people's hunting:

Many things lead to dilemmas. Sometimes I have spoken of Buddhist teachings more flexibly, and I have made a choice between indigenous people and our entire society. Then, I have had to pay the price for my choice. For example, to let indigenous people remain in the mountains, it may be better for them to stay there than to come down to the flatlands and go through all sorts of difficulties. This kind of thinking seems barbaric. Of course, there are still some other problems which may be connected. At a basic level, this sort of theory must be supported by an assumption. That is, people are more important than animals. The universe is made up of a hierarchy and animals are weaker. Therefore, relatively speaking, I do not tend to put humans first, but I am more of a realist. What I mean is that I do not keep too high of a profile. Sometimes I think that when we choose the path we want to walk on, we should act like the Buddha. Perhaps, you might sacrifice yourself by giving your life to the tiger. You may be willing to protect yourself and ignore the misery other living beings, or you may be willing to become an Arhat, but, whatever you do, do not harm other living beings. Moreover, if you are impatient and believe that you are just a normal person, you have to first appreciate your existence. Only then can your continued existence be assured. Furthermore, you can do the same, but you must have some limitations. You cannot harm other beings without any reason or cause. Also, can you use your ideology to offend other beings? Thus, a so-called ideology does not result from the singular choice of either you or me dying. You are very aware that there is a third path to follow, but still, you take harming animals lightly. Please tell me, what other reason do you have to harm them.

In the above quotation, Chao-hui uses the argument of dependent arising to demonstrate the idea that "all beings are equal." She makes it clear that her views are not based on her own personal perspective. Thus, the highest form of action is to give one's life to feed the tiger, just as the Buddha did. However, she also realizes that people today face the practical choice between helping themselves and helping others. Thus, she proposes a minimum requirement of at least not harming living beings without reason or cause. According to Chao-hui, both indigenous people and animals are vulnerable groups. Though she chooses to speak on behalf of animals, however, she continually points out that it is necessary to actively search for a "third path." As to where that path actually lies, she has yet to say! In summary, by looking at the process of amending the Wildlife Conservation Law, we can see a heated confrontation between the two groups. However, this confrontation was a product of the social context of the time. As to whether or not this confrontation is reconcilable, there is still space available within the discourses of both camps. It is this space that is most worthy of adoration.

Conclusion: Returning to J.B. Callicott's Ecological Holism

Though this study deals with two controversies that took place at different times, in different places with varying contexts, both controversies share a common background, that being people's attitudes toward animals. More specifically, both controversies have to do with the attitudes of indigenous people toward animals. This study began by looking at the debate sparked by J.B. Callicott's work, and then moved on to discuss the core "land ethic" values and the concept of ecological holism described by Callicott. Next, this study turned to the context of Taiwan to discuss the controversy taking place between the various discourses of the indigenous people's hunting, nature conservation and animal liberation (primarily based on Buddhist values) movements in the 1990s. Bringing the discussion up until now, the crux of the problem seems to lie in people's attitudes toward wild animals. The former controversy took place in academic journals. After a number of heated confrontations, the various positions were clarified, and all parties were able to respect and appreciate each other's stances. The latter controversy is not confined to the world of academia, but is closely related to the complicated social and political context of Taiwan. Even now, the controversy seems to be stuck in a stalemate, with no platform for dialogue available! The most recent debate took place at the end of 2004. This round of discussions was spurred by the "opening of the Danda forest road to Bunun hunting."[8] The idea was brought up by the Forestry Bureau. However, it came to an end before it even began. Nonetheless, the above analysis shows that, based on Leopold's concept of ecological holism, Callicott proposed some points that can potentially create room for dialogue between the "environmental justice" concept of the indigenous people's hunting movement, the conservationist theory of "wise use," and the discourse of "dependent arising and protecting life" espoused by the animal liberation movement. For example:

1. The animal liberation discourse in Taiwan is not firmly attached to the Western stances of animal rights or utilitarianism described above. Though the social actions of the Life Conservation Association have at times emphasized the ideas of Peter Singer, and Peter Singer himself has appeared in Taiwan on numerous occasions, from the context of the amendments of the Wildlife Conservation Law, we see that the anti-hunting position of the animal liberation movement in Taiwan is closely associated with the conservationist idea of "preservation," and Master Chao-hui's interpretations of the Buddhist teachings of dependent arising and "protecting life." Even more interestingly, there is a close connection between the Buddhist idea of dependent arising and the ecological principle that "everything is connected with everything else." Thus, the land ethic proposed by Leopold and promoted by Callicott can help to create even more room for dialogue.

2. As described by activist Taiban Sasala and scholar Chi Chun-chieh, indigenous people's hunting culture in Taiwan contains a wealth of ecological knowledge and standards. The matter of great concern is that these standards have been destroyed by the forces of modernization and they are disappearing at an alarming rate! The ecological holism proposed by Callicott is not merely concerned with animals; rather, it shows broader concern for the health of the ecosystem. Of course, this

includes respect for human social groups. Therefore, the land ethic should play a complementary role in the revitalization and rebuilding of indigenous people's hunting culture.

3. In the past ten years in Taiwan, proponents of the "wise use" model for the management of natural resources have fought hand in hand with the conservationists who promote "preservation." Members of these factions have gradually started to look for opportunities to present their management techniques and theories. For example, in Wutai Township, Pei Jia-chyi and Taiban Sasala drafted an experimental plan to manage wildlife. The plan aims to achieve the sustainable use of wildlife resources through organized management and wildlife monitoring techniques. These kinds of approaches offer greater space for development than simple prohibition policies. Moreover, Callicott emphasizes the well-being of the entire group rather than that of the individual. This can provide a solid ethical foundation for the "wise use" management model.

Based on the three points given above, it is possible for the indigenous people's hunting, conservation and animal liberation movements in Taiwan to come together. However, this sort of inference does not mean that such a result can necessarily be achieved at a social level. The actual results still depend on the creation of meaningful dialogue between members of the three camps. The series of debates sparked by Callicott and the important reminders given by Jamieson lead to the conclusion that an alliance between the factions can be formed. As Jamieson pointed out, all of the various discourses have arisen within social movements. Thus, after exploring the various social contexts from which they arose, we may discover that the differences that separate them are not as large as we once thought. However, the hunter and the hunted most definitely do have different interests. To bring this paper to a conclusion, a story told to the author by Tayal elder Masa Tohui shows that the difference between these interests may not be so great after all. The story goes something like this:

> The rainbow bridge story is an important traditional myth of the Tayal people. Honest and kind Tayal people are permitted to cross the bridge when they die to get to paradise on the other side. Those who are not cannot cross. Also, if the person is a hunter, then the spirits of all the animals they have killed cross the bridge with them. When they get to the other side, they can live forever.

Hearing this kind of myth from the mouth of an elder, the heart is filled with confusion! Isn't the relationship between the hunter and the hunted an antagonistic battle for life and death? Isn't it too romantic to believe that this antagonism is resolved at the time of death? Or does the Tayal view of hunting contain some ecological knowledge that is worthy of our attention? These questions conclude this study's efforts to search for a possible alliance between the indigenous people's hunting, conservation and animal liberation movements.

References

Berkes, F., and Folke, C., eds. 1998. Linking Social and Ecological Systems: Management Practices and Social Mechanisms for Building Resilience. Cambridge: Cambridge University Press.

Brush, S. B., and Stabinsky, D., eds. 1996. Valuing Local Knowledge: Indigenous People and Intellectual Property Rights. Washington, DC: Island Press.

Callicott, J. B. 1980. "Animal Liberation: A Triangular Affair." Environmental Ethics 2: 311-38.

Callicott, J. B. 1988. "Animal liberation and environmental ethics: Back together again." Between the Species 4: 163-69.

Callicott, J. B. 1989. In Defense of the Land Ethic: Essays in Environmental Philosophy. Albany: State University of New York Press.

Callicott, J. B. 1998. "Back Together Again." Environmental Values 7: 461-75.

Chang, L. S. 2001.〈論原住民的狩獵傳統與國家公園法的修正〉"On Amending National Park law Related to Indigenous Hunting Tradition." The Nature Quarterly, Society for Wildlife and Nature. 70：98-101。

Chen, M. T. 1973. "From Swidden to Orchard: Tautsa and Kamujiau's Adjustment to Agricultural Change." Bulletin of the Institute of Ethnology, Academia Sinica. 36：11-33。

Chen, S. L. 2002. "The Predicament of Indigenous Rights Legalization within Indigenous Movement: Discussions related to the Construction of Indigenous Autonomy". Master's Thesis, College of Law, National Taiwan University.

Chi, C. C. 1996. "Indigenous Hunting Culture from the Viewpoint of Environmental Justice". The Fourth World.〈http://wildmic. npust.edu.tw/sasala/index.htm〉。

Fu. C. 1997.〈台灣原住民「生態智慧」與野生動物保育〉"Taiwan's Indigenous Ecological Knowledge and Wildlife Conservation". Taiwan Indigenous Voice Bimonthly. 17：42-51。

Huang, Y. 2001. "Hunting Culture Without Hunters". Newsletter of Taiwan Environmental Information Center. http://news.ngo.org.tw/reviewer/huangyi/2001/re-huangyi01011001.htm

Huang, Y. G. 1993. "Crops, Economy and Society: A Case from Tongpu's Bunun People". Bulletin of the Institute of Ethnology, Academia Sinica. 75：133-69。

Jamieson, D. 1998. "Animal Liberation Is an Environmental Ethic." Environmental Values 7: 41-57.

Leopold, A. 1949. Sand County Almanac. Oxford: Oxford UP.

Lin, C. C. 1996.〈心淨則國土淨：關於佛教生態觀的思考與挑戰〉(Mind Purification and then Land Purification: Thought and Challenge of Buddhists' Ecological Viewpoint)。收於釋傳道《佛教與社會關懷：生命、生態、環境關懷論文集》。台南：現代佛教學會與中華佛教百科文獻基金會。179-92。

Lin, Y. R. 1992. "A Study on the Hunting Problems in Wutai District of Pingtung County of Taiwan." Unpublished Master Dissertation, University College London.

Lin, Y. R. 1998. Hunters Who Drink the Poisoned Wine: Science and Religion in Environmental Issues. Taipei: Meta Media.

Lin, Y. R. 1999. "The Environmental Beliefs and Practices of Taiwanese Buddhists." Unpublished PhD Dissertation, University College London.

Lin, Y. R. 2001.〈原住民部落發展與山林資源保育：從西雅圖酋長的演說談起〉"Indigenous Community Development and Forest Resources Management: A Viewpoint from the Speech of Chief Seattle". 發表於「保護區管理的國際新趨勢」研討會，2001 年 5 月 29、30 日。內政部營建署、中華民國國家公園學會、太魯閣國家公園管理處、台灣大學地理環境資源學系。台北、花蓮。

Lin, Y. R. 2006. "The Cultural Construction of Nature: Contesting the "Forest" of Proposed Maqaw National Park." Newsletter for Research on Applied Ethics 37: 7-23.

Lu, Y. 1995. "Reflecting Upon the Relationship Among the Doctrine of Mahayana, Environmental Protection and Social Reform" Con-temporary Monthly 107：130-41。

Midgley, M. 1983. Animals and Why They Matter. Athens: University of Georgia Press.

Naess, A. 1973. "The Shallow and the Deep, Long-Range Ecology Movement: A Summary." Inquiry 16: 95-100.

Nowell, K. 1991. "Aboriginal Hunting in the Tawu Mountain Nature Reserve Area, Taiwan." Unpublished Report to the Wildlife Ecology Laboratory, National Taiwan University, Taipei.

Passmore, J. 1974. Man's Responsibility for Nature: Ecological Problems and Western Traditions. New York: Scribner's.

Patel, A., and Lin, Y. S., 1989. History of Wildlife Conservation. Taipei: Council of Agriculture.

Rabinowitz, A., and Lee, L. L., 1990. "Management and Conservation Strategy for the Tawu Mountain Nature Reserve, Taiwan, R.O.C." Report to Council of Agriculture, Executive Yuan.

Regan, T. 1981. "The Nature and Possibility of an Environmental Ethic." Environmental Ethics 3: 19-34.

Regan, T. 1983. The Case for Animal Rights. Berkeley: University of California Press.

Reid, D. 2010. Nation versus tradition: A case study of indigenous rights. In David Blundell (Ed.). Taiwan Experience After Martial Law. University of California Press, Berkeley. (in press)

Routley, R. 1973. "Is There a Need for a New Environmental Ethic?" Proceedings of the Fifteenth World Congress of Philosophy. Ed. The Bulgarian Organizing Committee. Sophia, Bulgaria: Sophia Press.

Sagoff, M. 1984. "Animal Liberation and Environmental Ethics: Bad Marriage, Quick Divorce. Osgoode Hall Law Journal 22: 306-32.

Singer, P. 1977. Animal Liberation: A New Ethics for Our Treatment of Animals. New York: Avon.

Taiban Sasala. 2000. "Seeking the Lost Arrows: Thought of Indigenous People's Hunting." The Fourth World. 〈http://wildmic.npust.edu.tw/sasala/尋找失落的箭矢.htm〉

Taiban Sasala. 2002. "The Myth and Facts of Indigenous Hunting Culture: A Political Ecological Observation." 《南方電子報》〈http://www.esouth.org/sccid/south/south20020228.htm〉。

Taylor, P. W. 1986. Respect for Nature: A Theory of Environmental Ethics. Princeton: Princeton University Press.

Wang, M. H. 2001. "A Reflection and Prospect of Taiwan's Indigenous People's Movement" National Taiwan Normal University Department of Geography. 〈 http://www.geo.ntnu.edu.tw/faculty/tibu/台灣原住民族運動的回顧與展望發表稿.htm 〉

Wang, Y. 1986. "A Wildlife Resource Utilization Survey on Game Meat Shops in Taiwan (I) Council of Agriculture, Executive Yuan, Ecological Research-011.

Wang, Y. 1988. "A Wildlife Resource Utilization Survey on Game Meat Shops in Taiwan (III)". Council of Agriculture, Executive Yuan, Ecological Research-017.

Wang, Y., and W. C. Lin. 1987. "A Wildlife Resource Utilization Survey on Game Meat Shops in Taiwan (II)". Council of Agriculture, Executive Yuan, Ecological Research-021.

Warren, M. A. 1983. "The Rights of the Non-Human World." Environmental Philosophy: A Collection of Readings. Ed. R. Elliot and A. Gare. University Park: Pennsylvania State University Press.

Yaffee, S. L. 1994. The Wisdom of the Spotted Owl: Policy Lessons for a New Century. Washington, D. C.: Island Press.

Yang, H. N. 1994. ＜當代台灣佛教環保理念的省思：以「預約人間淨土」和「心靈環保」為例＞ (A Review of Contemporary Taiwanese Buddhist Huan-Bao Beliefs: Cases from Prescribing Pureland on Earth and Spiritual Huan-Bao), Con-temporary Monthly 104：32-55。

Endnotes

1. The term "animal liberation" is primarily used to refer to the activities of the Life Conservationist Association (LCA) (Section 2 includes a discussion of this organization). This, of course, is not the only organization in Taiwan concerned with the treatment of animals, but in the 1990s, the LCA brought issues related to animals to the attention of society. Moreover, through its practical work, this organization has

made the orientation of its social movement very clear. Outside Taiwan, organizations concerned with animals have different approaches. Some groups, like the Royal Society for the Prevention of Cruelty to Animals (RSPCA), base their movements on humanitarian concerns for animal welfare. Other groups advocate more radical stances based on animal rights. In fact, various organizations have a wide range of approaches. The term "liberation" is derived from the title of a book by philosopher Peter Singer first published in 1977. His position is not based on the idea of rights. Rather, he surpasses the traditional humanitarian position with a stance that is more similar to that of the LCA. That is, he does not advocate animal rights and does not take the traditional humanitarian stance. Accordingly, this study uses the phrase "animal liberation movement," which implies the desire for social reform. Nevertheless, it is important to distinguish between the stance of the LCA, and the philosophical stance of utilitarianism taken by Singer.

2. The Maqaw National Park controversy arose due to the efforts of conservationists to protect old-growth cypress forests. Starting in 1998, conservationists began a social movement calling upon the government to set up reserves to protect precious old-growth forests. Since the contested regions fall within the traditional territory of the Tayal people, the movement attracted the concern of indigenous groups. As a new government administration took control in 2000, conservation organizations and some indigenous groups worked together to promote the idea of the indigenous management of a new type of national park. The plan was later abandoned due to the objections of other indigenous groups. For a more detailed description of this controversy, see Lin (2006).

3. The wind-fallen beech tree event in Smangus took place in 2005. After a community meeting, three members of the Smangus community collected beech trees that had previously been blown over in a typhoon. They were then detained and charged with theft by the Forestry Bureau. Members of the Smangus community objected to the charges on the grounds that the community members acted in accordance with traditional practices of exploiting resources within their traditional territory. Based on the Indigenous Peoples' Basic Law and the claim that no criminal offense had been committed, community members

repeatedly appealed the criminal charges. Finally, in early 2010, the High Court pronounced the plaintiffs not guilty. This case made history by being the first case in which the Taiwanese legal system made a verdict based on traditional indigenous practices, See Reid (2010).

4. The changes in the National Park Law prompted by the Maqaw and Smangus controversies led to amendments of laws related to indigenous people's hunting by the Executive Yuan. For example, through consultation with the National Park Service and authorization from the Ministry of the Interior, indigenous and other residents living within the boundaries of national parks may be exempt from regulations prohibiting certain types of activities, including burning of vegetation and pieces of land, hunting animals and catching fish, and collecting plants and wood.

5. Similarly, there was an intense debate between Arne Naess's deep ecology and so-called "social ecology." The leading proponent of social ecology was Murray Bookchin. Jamieson does not mention this debate. However, it is also revealing of the wide range of factions within the environmental movement.

6. The main for reason this is that Taiwan still lacked the technology to monitor large and midsized mammal species. As a result, scholars were not able to understand actual trends and changes of wildlife populations in mountain regions. Research and development of monitoring technology in Taiwan, such as wireless tracking and automatic cameras, was not widely used until the mid-1990s. Only then did highly reliable information become available.

7. The rhinoceros horn event occurred at the end of 1993. The English Environmental Investigation Agency (EIA) identified Taiwan, Hong Kong and China as the world's leading consumers of rhinoceros horns. By connecting international environmental groups and utilizing the media, environmentalists attempted to sanction the Taiwanese wild animal trade at CITES, the important annual international environmental conference held in Washington D.C. Though the effort was not successful, President Clinton signed the Pelly Amendment Act banning the export of American wild animal products to Taiwan. This series of international environmental actions put a significant amount of pressure on the Ministry of Foreign Affairs and the Council of

Agriculture. Due to the pressure from Congress, the Council of Agriculture amended the Wildlife Conservation Law in the hopes of discouraging the illegal rhinoceros trade by raising fines. Though this event led to legal amendments, the range covered by the amended law was not limited to rhinoceros horns. During the amendment process, the LCA proposed their views on animal liberation, which differed from conservationists who advocated allowing hunting on a conditional basis.

8. From December 14-23, 2004, the Forestry Bureau selected the Danda region (within the Bunan sphere of livelihood) to try out the, "Danda regional trial hunting plan to meet the cultural and ceremonial needs of indigenous people," in order to experiment with the possibility of combining community participation, conservation and indigenous people's hunting indigenous people's hunting culture. Reports of the plan in the media started another round of intense debates. For more details, see the "Hunting and Conservation Forum" on the Environmental Information Center's website.

CHAPTER FOURTEEN

Taiwan's Reform of Energy Law and Policy for Mitigation of Climate Change

Jui-Chu Lin, Tsung-Tang Lee

I. Introduction

Climate change mitigation has been a fundamental consideration in the formulation of contemporary environmental policies. According to the Intergovernmental Panel on Climate Change (hereafter referred to as the 'IPCC'), changes in precipitation and temperature will lead to changes in runoff and water availability. This will exacerbate the effects on water resources that already under stress from population growth and urbanization.[1] Increases in the frequency and severity of floods and droughts have been projected in the Fourth Assessment Report (hereafter referred to as the 'AR4'), which noted the adverse impacts of climate change on biological ecosystems and human communities around the world.[2] In Taiwan, the connection between the rise in global temperature and increased risks of floods and droughts has been proven by a research team composed of academics from Taiwan and mainland China,[3] with the risk of floods being realized by a devastating typhoon in August 2009. The mitigation of the causes of climate change, therefore, is essential in the goal of protecting the environment on which human societies rely for sustainable development and lasting survival.

In order to reduce the concentration of greenhouse gases (hereafter referred to as 'GHG') in the atmosphere, which are identified as the principal cause of global warming, decreases in GHG emissions resulting from the consumption of fossil fuels are therefore much desired by policymakers. As pointed out by the AR4, "global GHG emissions due to human activities have grown since pre-industrial times, with an increase of 70% between 1970 and 2004." Furthermore, CO_2 emissions from fossil fuel use accounted for 56.6% of total global GHG emissions in 2004.[4] It is therefore widely believed that energy efficiencies and the substitution of renewable energy technologies, which reduce the amount of energy consumed and GHG emitted from energy consumption, will help mitigate the causes of climate change once sufficiently diffused.

This essay analyses energy law and policy from an economic perspective. As noted by the Stern Review, "the underlying ethics of basic welfare economics, which underpins much of the standard analysis of public policy, focuses on the consequences of policy for the consumption of goods and services by individuals in a community."[5] Policymaking in the context of climate change involves ethical judgments inspired by moral notions regarding justice and equality among different nations and generations. The analysis of energy law in such circumstance attracts interest in the nature of the law. Legal philosopher Neil MacCormick had stated: "Law is both a normative and an institutional order, and this connects it with the two poles of the contrast. As a normative order, it replicates certain features of morality, and connects necessarily to morality in certain ways. As an institutional order, it connects necessarily to politics and is in part constitutive of the political, while well-conducted politics is necessary to the maintenance of satisfactory systems of positive law, especially state-law."[6] The influence of environmental ethics has been indispensable in forming the present normative order that mandates lawmakers to carry out a reform of energy law.

Although Taiwan is not a signatory of the United Nations Framework Convention on Climate Change (hereafter referred to as the 'UNFCCC'),[7] as a stakeholder on climate change, Taiwan has committed itself to the reduction of GHG emissions. Taiwan has introduced a series of legislation aimed at both setting new standards of energy efficiency and creating a market for renewable energy.

The purpose of this chapter is fourfold. (1) To outline the GHG profile and energy supply structure of Taiwan; (2) to investigate Taiwan's GHG reduction targets and its voluntary nature; (3) to discuss the rationale of green energy policy from an economic perspective; and (4) to look at Taiwan's energy efficiency and renewable energy legislation.

II. GHG Emissions and Energy Supplies in Taiwan

Currently, Taiwan ranks one of the world's 20 largest economies in terms of GDP and one of the top 20 GHG emitters in per capita terms. This section examines the relationship between Taiwan's GHG emissions and its dependence on imported fossil energy.

1. Taiwan: One of the Largest Emitters

GHG emissions from Taiwan have been large since the country's industrialization, especially in per capita terms. According to the International Energy Agency (hereafter referred to as the "IEA"), Taiwan (identified as Chinese Taipei) had 276.18 Mt of CO_2 emissions from fuel combustion in 2007. With a population of 22.86 million, CO_2 emissions per capita of Taiwan was 12.08 tons. This was significantly higher than that of other industrialized East Asian economies such as Japan (9.68 tons) and South Korea (10.09 tons). This result ranked Taiwan 18th compared to the rest of the world.[8]

Moreover, the GHG growth rate of Taiwan could be highest among those of all major economies. Statistics published by the Environmental Protection Administration (hereafter referred to as the 'EPA') of Taiwan revealed that GHG emissions had increased by 124.68% during the period between 1990 and 2006, from 123.574 Mt carbon dioxide equivalent (hereafter the 'CDE') to 277.645 Mt CDE. The increase in energy consumption was identified as the principal force driving the growth of total GHG emissions in Taiwan. As a result, the carbon dioxide emitted by the energy sector accounted for 95.5% of Taiwan's overall CO_2 emissions in 2006.[9]

2. Dependence on Imported Fossil Energy

In 1876 under the rule of the Chinese Qing Dynasty, Taiwan opened its first large-scale coalfield near the northeastern city of Keelung. Since then, Taiwan's coal mining industry has been prosperous and at its height counted some 382 mine fields across the island. In 1960, the Taiwanese mining industry extracted approximately four million tons of coal with much of the total being used for export. Throughout the 1960s Taiwan's energy consumption gradually shifted to imported petroleum however. With the shift the demand for coal declined dramatically and local coal mining costs rose. As a result, the country's coal industry could no longer compete with other types of energy or even with coal imported from overseas. The 1990s saw the final collapse of the Taiwanese coal industry. In 2000, Taiwan closed its remaining four coalfields.

Today, Taiwan relies on imported energy to fuel its economy. According to the Bureau of Energy, imported energy accounted for 99.3% of energy supply in 2008. This included 91.3% from fossil fuels such as petroleum and coal and the remaining 8% derived from nuclear sources. Renewable energies such as solar power, solar heat, wind power, and water power generated a regrettable 0.6% of Taiwan's total energy needs in 2008.[10] It is, therefore, evident that Taiwan depends on imported fossil fuels for the majority of its energy needs.

As energy consumption accounts for more than 95% of CO_2 emissions from Taiwan, and more than 91% of energy supplied in Taiwan is generated from fossil fuels, it is obvious that the only way to reduce Taiwan's GHG emissions on a meaningful scale is to reform the country's energy sector. At this juncture it should be pointed out that although it is widely believed that the GHG emission footprint of nuclear power plants are diminishingly small compared to those of fossil fuel-fired power plants, the environmental hazards posed by nuclear waste have made nuclear power politically unpopular in Taiwan. Consequently, renewable energy is the most likely substitute for fossil fuels. Therefore, in the case of Taiwan, it is safe to say that the success of GHG emissions reduction depends on the sufficient exploitation of renewable energy resources.

III. Taiwan's Voluntary Emissions Reduction Target

While not being a signatory to the Kyoto Protocol,[11] Taiwan has voluntarily adopted an emissions reduction target. This section looks into the relationship between Taiwan and the UNFCCC regime, Taiwan's voluntary emissions reduction target, and issues regarding the draft Greenhouse Gas Reduction Bill.

1. Taiwan and the UNFCCC

Formally known as the Republic of China, Taiwan represented China in the United Nations until October 1971, when the General Assembly recognized the representative of the People's Republic of China as, "the only lawful representative of China to the United Nations."[12] Since then, Taiwan has been placed outside the United Nations system – including the UNFCCC.

That is to say, the UNFCCC and the Kyoto Protocol do not officially bind Taiwan. As a result, the political momentum towards a mitigation of the causes of climate change remained weak in Taiwan until extreme weather brought public attention to the issue in recent years. For instance, the Executive Yuan proposed the draft Renewable Energy Bill to the Legislative Yuan in 2002, but the Bill was not converted into the Renewable Energy Act until seven years later in June 2009. Another example can be found in the handling of the draft Greenhouse Gas Reduction Bill, proposed to the Legislative Yuan for the first time in September 2006 and, at the time of writing, still unable to garner sufficient support for its passage.

2. Emissions Reduction Target of Taiwan

Taiwan has set its voluntary emissions reduction target as part of the 'Guiding Principles of Sustainable Energy Policy.' This was approved by the Executive Yuan in June 2008 and is aimed at returning the country's CO2 emissions to 2008 levels during the period between 2016 and 2020, and to 2000 levels by 2025.[13]

The 'Guiding Principles' mandates that the government promote clean energy in the energy sector and energy efficiency in other sectors of the society. In order to provide the legal basis of green energy policy, the 'Guiding Principles' commits the government to push for the enactment of the draft Greenhouse Gas Reduction Bill, the draft Energy Tax Bill, the draft Renewable Energy Development Bill (now the Renewable Energy Development Act), and the draft Energy Management Amendment Bill (now the Amendments of the Energy Management Act).

It is noteworthy that, although the reduction target has already been set by the 'Guiding Principles,' questions regarding the government's competence in both the setting of emissions reduction targets and the timetables for implementation have been the most debated during the parliamentary review of the draft Greenhouse Gas Reduction Bill. Accordingly, the reduction target set by 'Guiding Principles' should probably be best seen as a political rather than binding commitment.

3. The Draft Greenhouse Gas Reduction Bill

The draft Greenhouse Gas Reduction Bill was proposed to the Legislative Yuan for the first time in September 2006, and has since been revised several times by the Executive Yuan and by individual Legislators. Apart from the questions of competence to set reduction targets (Article 13), the emissions trading mechanism, outlined in Article 15, is arguably the most essential and controversial clause of the Bill.

In its cap-and-trade program, the Bill, in principle, adopts an 'absolute cap,' designed to impose a physically quantified emissions cap during a specific period of time, and to allow participants who emit less than their quotas to sell the residual allowances to those whose actual emissions exceed the cap imposed on them. Opposite to the 'absolute cap' model is the 'relative cap,' which requires participants to meet certain reduction rates. Internationally, Canada and the Australian state of New South Wales offer examples of states or territories that have adopted the 'relative cap' model for their emissions trading schemes.[14]

The Bill, on the other hand, provides an exception to the 'absolute cap' model. In order to encourage companies to undertake emission reductions before the national target is set under the legislation, Article 16 of the Bill allows such companies to exchange amounts of reduction – certified by designated institutions and validated by the competent authority – in return for reduction credits that may be traded or used to offset the companies' emissions after the formal introduction of the national reduction target. For example, the Taiwan Power Company (hereafter referred to as the 'TPC'), a state-owned electricity monopoly responsible for 51.13% of corporate GHG emissions in Taiwan,[15] has launched 7 reduction programs under the 'Project of Greenhouse Gas Reduction in the Energy Industry' run by the Energy Bureau and reduced some 69,000 tons of CO_2 emitted since 2006. The TPC has been hoping that the amount of GHG reduction achieved since 2006 may be exchanged for reduction credits, which, after the introduction of the national target, can offset its excessive emissions in the cap-and-trade mechanism under the Bill.

Despite such incentives, support for the Bill has been insignificant. Corporations have argued that the 'cap' aspect of the emissions trading mechanism would weaken Taiwan's economic competitiveness, while environmentalists have been skeptical of the 'trade' portion. In June 2009, when the Legislative Yuan enacted the Renewable Energy Development Act and the Amendments of the Energy Management Act, the draft Greenhouse Gas Reduction Bill, despite being part of the Green Energy Package, did not attract consensus among the Legislators.

IV. The Rationale of Green Energy Policies

As noted earlier in this chapter, energy produced by the burning of fossil fuels accounts for the majority of the GHG emissions from Taiwan. The success of emissions reduction, therefore, depends largely on the diffusion of green energy technologies. This section discusses the rationale of policies aimed at promoting green energy and how they relate with other emissions reduction policies.

1. Environmental Externality and Unpriced Public Goods

The principal objective of a green energy policy should be to correct the market failure caused by the environmental externality costs of fossil fuels and the unpriced public goods generated from the use of green energies. In the context of this essay, the term 'environmental externality costs' refers to the environmental costs of the burning of fossil fuels that are born by the society as a whole, while 'unpriced public goods' refers to the environmental benefits generated from the use of green energy that is not valued by the pricing mechanism of the market.

As defined by economists, "external costs exist when the private calculation of benefits or costs differs from society's valuation of benefits or costs."[16] In the case of fossil fuels, the damage that climate change may bring about is born by the human community as a whole, while the consumers directly or indirectly responsible for fossil fuel GHG emissions are not charged for such societal costs. As a result, the prices of fossil fuels and electricity generated from fossil fuel combustion are much lower than might otherwise be the case.

On the other hand, the environmental benefits that green energy can generate with its potential for reducing GHG emissions from energy consumption are public goods, which spill over to the overall community but are not otherwise appreciated in the market. Consequently, prices in a market without government intervention may not reflect the societal advantages of green energies, making them less attractive to investors than they would otherwise be under optimal conditions.

As noted by Marilyn A. Brown, in the case of green energy, environmental externality costs and unpriced public goods are the two prime causes of market failure: (1) Fossil energy using contemporary conversion technologies produces a variety of unpriced costs (or negative externalities) including greenhouse gas emissions, as a result of these unpriced costs, more fossil energy is consumed than is socially optimal;[17] (2) Because public goods are unpriced, markets tend to under produce them. For instance, although benefits might accrue to society at large, individual firms cannot realize the full economic benefits of their R&D investments in green technologies. The payback from advances in green energy technologies is not only experienced by the sponsoring company, but also flows to the company's competitors and to other parts of the economy.[18]

2. Correcting Market Failures

In order to correct market failure that impedes the diffusion of green energy technologies, governments around the world have introduced a variety of policy measures aimed at internalizing the environmental externality costs of fossil fuels and at appreciating the environmental benefits of green energies in the marketplace.

In theory, the most straightforward method used to internalize environmental externalities and to implement the 'polluter pays' principle in the energy sector is to introduce a GHG tax – although the practicalities of its implementation would be fairly complicated.[19] As explained by Anthony D. Owen, "the tax would be required to be imposed at differential rates, depending upon the total estimated damages resulting from the fuel in question."[20] Nevertheless, taking into consideration the impacts on lower-income communities and trade-related sectors, national and international consensus must be secured before such GHG tax is introduced.

Aside from a direct GHG tax it is also true that incentive measures may be a significant method of appreciating environmental benefits and may gently encourage the development of green energy. By subsidizing R&D activities, commercial applications, and the procurement of green energy products and services, incentive measures can help reduce the costs of green energy development as well as boost demand for green energy. It is also true however, that the government may not always be best placed to make decisions on the specific choices of green energy technologies for each geographic community. Policy-makers of green energy policies, while attempting to correct market failures, should not undermine market mechanisms altogether.

A market-based alternative for both the internalization of environmental externalities and the appreciation of environmental public goods is 'Emissions Trading.' Emissions trading schemes allow participants to trade surplus carbon credits with each other. This policy option would reward companies that invest in green energy facilities and penalize those who do not. In other words, emission trading schemes internalize the environmental externality costs by forcing businesses that emit more than their quota to purchase further carbon credits. It also appreciates the environmental benefits achieved by more efficient companies by allowing them to trade their surplus carbon credits – thus making their businesses more profitable.

As put forward by academics at the Oak Ridge National Laboratory, there are a range of solutions available to policymakers to achieve the correction of market failures which include:[21] (1) Carbon trading system; (2) Increased R&D resources; (3) Portfolio standards and production tax credits for renewable energy; (4) Voluntary fuel economy agreements with auto manufacturers in the transportation industry; and (5) Energy efficiency standards and labeling programs of buildings.

V. Green Energy Legislation in Taiwan

Bound by the constitutional principle of *Rechtsstaat*, or 'state of law' in which the government shall exercise its power in accordance with the law, explicit authorization of the law is needed before the Taiwanese government may make a policy which involves a 'significant matter.' Green energy policies, that restrict energy consumption choices or decide on the distribution of 'significant' public funds must therefore be made in accordance with Acts of Parliament.

1. The Principle of *Rechtsstaat* in Taiwan

After Taiwan's democratization in late 1980s and early 1990s, *Rechtsstaat* has been one of the most invoked principles in interpretations of the Constitutional Court. In its Judicial Yuan Interpretation No. 614, the Constitutional Court ruled that: "The modern principle of a constitutional state is specifically manifested by the principle of legal reservation under the Constitution. Not only does it regulate the relations between the state and the people, but it also involves the division of powers and authorities between the executive and legislative branches....If...any **significant matter** [emphasis added] is involved, e.g., public interests or protection of fundamental rights of the people, the competent authority, in principle, should not formulate and issue any regulation without express authorization of the law (see J. Y. Interpretation No. 443)."[22]

A policy aimed at promoting green energy therefore, which may involve matters such as: the transformation of the energy sector, protection of the environment, restrictions on the choice of energy-related products and services, or, the distribution of public funds, must be founded on an Act of Parliament insofar as its involvement in these matters is an exercise of governmental power. The Amendments of the Energy Management Act and the Renewable Energy Development Act were enacted to provide just such a legal basis for energy efficiency and renewable energy policies.

2. The Amendments of the Energy Management Act

The Energy Management Act was enacted in 1970, and has been amended multiple times since then. In July 2009, as part of the Green Energy Package, several amendments of the Act were enacted by the Legislative Yuan. Among the amendments, Articles

14 and 15 are designed to provide the legal basis for energy efficiency policies.

Article 14 regulates the energy efficiency of energy utilization facilities and apparatus – apart from those categorized as automobiles. Article 14(1) obligates manufacturers and importers of energy utilization facilities and apparatus that are designated by the central competent authority to comply with energy consumption standards set by the central competent authority and to label their products with energy consumption and efficiency information, if those products are manufactured or imported for use in Taiwan. Article 14(2) prohibits energy utilization facilities and apparatus that fail to conform to the energy consumption standards from being imported to or sold in Taiwan. Article 14(3) forbids the display and sale of energy utilization facilities and apparatus not labeled with energy consumption and efficiency information.

Article 15 (the wording of which is almost identical to that of Article 14) regulates the energy efficiency of automobiles. Article 15(1) obliges domestic automotive manufacturers and importers of designated automobiles to comply with the energy consumption standards set by the central competent authority and to label their vehicles with energy consumption and efficiency information. Article 15(2) prohibits automobiles that fail to conform to the energy consumption standards from being sold in or even imported into the Taiwanese market. Article 15(3) forbids the display and sale of automobiles not labeled with energy consumption and efficiency information.

3. The Renewable Energy Development Act

The Renewable Energy Development Act was enacted in July 2009 as part of the Green Energy Package. Article 1 indicates that the purpose of the Act is to promote the exploitation of renewable energy, enrich energy diversity, improve quality of the environment, advance relevant industries, and to secure the sustainability of national development.

(3.1) Renewable energy development fund

Article 7(1) of the Act authorizes the government to create a Renewable Energy Development Fund financed by funds collected from electricity suppliers who generate electricity from non-renewable sources and by government budget. Article 7(4) outlines that the purposes of the Fund are: (a) renewable energy electricity subsidies; (b) subsidy for renewable energy facilities; (c) projects for demonstration and the promotion of renewable energy; and (d) other renewable energy-related purposes approved by the central competent authority.

(3.2) Procurement of renewable energy electricity

Market access for electricity generated from renewable energy sources is the most essential issue dealt with by the Act. The provisions of Article 8 are aimed at removing market barriers to the capitalization and consumption of green electricity. Article 8(1) obliges electric utilities to give grid access to and to procure the electricity from renewable sources within their operation areas, unless a denial to such procurement is approved by the central competent authority. Article 8(1) also allows the central competent authority to appoint another electric utility to carry out the procurement if necessary. Article 8(2) provides that the costs of extending an electrical network in order to procure electricity from renewable source shall be shared by the electric utility and the owner of renewable generator. Article 8(3) requires that electric utilities, in procuring renewable electricity, shall sign contracts with owners of renewable generators and notify the central competent authority of the contracts.

The Act has also established a mechanism by which feed-in tariff rates shall be decided. In order to correct market failure mentioned above, Article 9 was designed to appreciate the environmental benefits of electricity from renewable sources. Article 9(3) provides that, in order to promote investment in renewable power generators, feed-in tariff rates shall not be lower than the average cost of electricity generated from fossil fuels. According to Article 9(1), feed-in tariff rates and the formula that determines the rates shall be decided by a committee composed of officials, experts, and concerned groups. The rates and the formula shall be reviewed annually, taking into account advances in technology, changes in cost, fulfillment of the target, and other relevant factors.

(3.3) Renewable heat incentive

Article 13 provides the legal authorization for incentive measures aiming at promoting the utilization of renewable heat. Article 13(1) mandates the government to formulate regulations that provide incentives to the exploitations of renewable heat such as solar heat and bio-fuels. Article 13(2) authorizes the government to use the Petroleum Fund provided in the Petroleum Management Act to subsidize exploitations that are substitutes for petroleum. According to Article 34(1) of the Petroleum Management Act, funds of the Petroleum Fund are collected from petroleum-related businesses.

VI. Conclusions

By enacting the Renewable Energy Development Act and the amendments of the Energy Management Act, Taiwan has established the legal foundation for green energy policies – namely programs aimed at promoting renewable energy and energy efficiency. Despite the passing of these pieces of legislation Taiwan is still in need of a market-based solution for the internalization of environmental externality costs and for the appreciation of environmental public goods in the energy sector. The emissions trading scheme provided in the draft Greenhouse Gas Reduction Bill, therefore, is crucial to the success of the 'green revolution' in the Taiwanese energy market.

Public support is essential to the reform of energy law and policy. Greater understanding of environmental ethics would help raise awareness of climate responsibility in the public arena, where policy debates take place. The mitigation of the causes of climate change can be achieved only if the environment is prioritized over short-term economic growth. On the other hand, given the declines in oil reserves around the world, green energy is the most feasible solution for sustainable economic growth in the long term. Therefore, it is fair to say that green energy policies, if appropriately formulated, would encourage investment in the environmental as well as economic future of mankind.

Bibliography

1. Alexander Rosnagel, 'Evaluating Links between Emissions Trading Schemes: An Analytical Framework' (2008) 4 CCRL 394.
2. Anthony D. Owen, 'Renewable Energy: Externality Costs as Market Barriers' (2006) 34 Energy Policy 632.
3. Dai-Gee Shaw, Lih-Chyi Wen, and Yung-Shuen Shen, *The Study of Building Carbon Emissions Trading Scheme* (CEPD, Taipei 2009).
4. IPCC, 2007: Climate Change 2007: Synthesis Report. Contribution of Working Groups I, II and III to the Fourth Assessment Report of the Intergovernmental Panel on Climate Change. IPCC, Geneva, Switzerland.
5. Liu, S. C., C. Fu, C.-J. Shiu, J.-P. Chen, and F. Wu (2009), Temperature Dependence of Global Precipitation Extremes, Geophys. Res. Lett., 36, L17702, doi:10.1029/2009GL040218.
6. Marilyn A. Brown, 'Market Failures and Barriers as a Basis for Clean Energy Policies' (2001) 29 Energy Policy 1197.
7. Marilyn A. Brown, Mark D. Levine, Walter Short and Jonathan G. Koomey, 'Scenarios for a Clean Energy Future' (2001) 29 Energy Policy 1179.
8. Nicholas Stern, *The Economics of Climate Change: The Stern Review* (CUP, Cambridge 2007).
9. Paul Krugman and Robin Wells, *Economics* (2nd edn Worth Publishers, New York 2009).

Notes

[1] IPCC, 2007: Climate Change 2007: Synthesis Report. Contribution of Working Groups I, II and III to the Fourth Assessment Report of the Intergovernmental Panel on Climate Change. IPCC, Geneva, Switzerland, at 49.
[2] *Ibid.*

[3] Liu, S. C., C. Fu, C.-J. Shiu, J.-P. Chen, and F. Wu (2009), Temperature Dependence of Global Precipitation Extremes, Geophys. Res. Lett., 36, L17702, doi:10.1029/2009GL040218.
[4] See note 1 above, at 36.
[5] Nicholas Stern, *The Economics of Climate Change: The Stern Review* (CUP, Cambridge 2007) 31.
[6] Neil MacCormick, *Questioning Sovereignty* (OUP, Oxford 1999) 15.
[7] United Nations Framework Convention on Climate Change (adopted 9 May 1992, entered into force 21 March 1994) 1771 UNTS 107 (UNFCCC).
[8] IEA, 'Statistics and Balances' <http://www.iea.org/stats/index.asp> accessed 23 February 2010.
[9] EPA, 'Statistics of GHG Emissions' <http://www.epa.gov.tw/ch/artshow.aspx?busin=12379&art=2009011715443552&path=12437> accessed 23 February 2010.
[10] Bureau of Energy, 'Energy Statistical Annual Reports' (Energy Supply and Consumption Flowchart 2008) <http://www.moeaboe.gov.tw/opengovinfo/Plan/all/energy_year/main_en/EnEnergyYearMain.aspx?pageid=EnEnergyYear> accessed 23 February 2010.
[11] Kyoto Protocol to the United Nations Framework Convention on Climate Change (adopted 11 December 1997, entered into force 16 February 2005) (1998) 37 ILM 22.
[12] UNGA Res 2758 (XXVI) (25 October 1971).
[13] GIO, 'Guiding Principles of Sustainable Energy Policy' < http://info.gio.gov.tw//ct.asp?xItem=37060&ctNode=3764&mp=1> accessed 24 February 2010.
[14] Alexander Rosnagel, 'Evaluating Links between Emissions Trading Schemes: An Analytical Framework' (2008) 4 CCRL 394, 400.
[15] Dai-Gee Shaw, Lih-Chyi Wen, and Yung-Shuen Shen, *The Study of Building Carbon Emissions Trading Scheme* (CEPD, Taipei 2009).
[16] Jonathan Koomey and Florentin Krause, 'Introduction to Social Externality Costs' in Frank Kreith and Ronald E. West (eds), *CRC Handbook of Energy Efficiency* (CRC Press, 1997) 153, 153.

[17] Marilyn A. Brown, 'Market Failures and Barriers as a Basis for Clean Energy Policies' (2001) 29 Energy Policy 1197, 1200.
[18] *Ibid*, 1201.
[19] Anthony D. Owen, 'Renewable Energy: Externality Costs as Market Barriers' (2006) 34 Energy Policy 632, 638.
[20] *Ibid*.
[21] Marilyn A. Brown, Mark D. Levine, Walter Short and Jonathan G. Koomey, 'Scenarios for a Clean Energy Future' (2001) 29 Energy Policy 1179, 1180-1182.
[22] J.Y. Interpretation No. 614 [2006] Judicial Yuan Report No. 10 Volume 48, 284-302.

CHAPTER FIFTEEN

How Students Conceptualize the Environment: Implications for Science and Environmental Education

Shiang-Yao Liu

This article reports an empirical study conducted to investigate students' conceptions of environment and their perspectives on the quality of the environment around them. It is based on the assumption that peoples' ideas about the environment may guide the ways in which they perceive environmental issues and further influence their willingness to take environmental responsibility. Differences in conceptions of environment among grade levels and community settings could be explained by students' life experiences, education and culture. Research on students' thinking about the environment may provide references for the development of educational programs.

Introduction

Over the past few decades, interest in and awareness of environmental issues have increased. Implementing environmental education in schools is viewed as an important strategy to broaden and deepen this awareness among people. That is to say, environmental education programs should be designed to raise students' concerns about environmental issues and foster their skills to solve the problems. However, not all of the environmental education programs are effective, it must embrace the "reflexive perspective" that both teacher and student bring knowledge to the environmental issue investigation process (Gauthier, Guilbert, & Pelletier, 1997). It is evident that learners do not come to educational programs with a blank-slate mind; they have their prior beliefs and values relating to the subject (Robertson, 1993). Their "views" or "preconceptions" should be explored in order to prepare a suitable program.

Understanding underlying beliefs and mental models of individuals, therefore, becomes an important research topic in the field of cognitive sciences because this understanding offers a practical reference to develop training programs or to resolve conflicts among target populations (Mueller & Veinott, 2008; Shutters & Cutts, 2008). In addition, research evidence shows that mental models and associated values are pertinent to environmental decision-making and further affect environmental behaviors (Atran, Medin, & Ross, 2005). Furthermore, as stated by Daugstad, Rønningen, and Skar (2006), conceptualizations serve as "markers of a position" and "legitimizing tools" playing a central role in the management agenda relating to environmental conservation. It is clear that environmental ideas can guide how people perceive environmental issues and further influence their willingness to take steps to be environmentally responsible.

In our society, diverging views of nature and the environment are manifest in peoples' everyday discourses and in a variety of social contexts. Assuming that, "people's behavior toward the natural world is surely conditioned in part by their ways of knowing and modeling it" (Atran & Medin, 2008, p.3), studies on how people conceptualize the natural environment are encouraged. Atran and his colleagues (2005, 2008), by conducting cross-cultural research, found that mental models of the natural environment differ among cultural groups living in the same area and engaging in similar activities. Their studies focused on the cultural consensus with respect to environmental ideas and considered the collective cognition of cultural groups in Mesoamerica and North America. They further inferred that students' conceptions and conceptual frameworks of the biological world may facilitate or interfere with science classroom learning.

An article by Lijmbach, Margadant-Van Arcken, Van Koppen and Wals (2002), entitled as 'Your View of Nature is not Mine', suggested that acknowledging differences in views of the environment among different target groups may provide information for educational programs in which environmental controversies with intergroup conflict over resource use are involved. Lijmbach *et al.* also recommended a shift in environmental education, from teaching people the "right" attitude, knowledge, and skills for environmental protection towards providing people opportunities to reflect on their own ideas and values pertaining to environmental issues, discuss with others, and develop collective perspectives that may guide one's behaviors and actions.

A number of recent studies have focused on young children's conceptions of the environment (Alerby, 2000; Littledyke, 2004; Loughland, Reid, & Petocz, 2002; Shepardson, Wee, Priddy, & Harbor, 2007). Each study developed its own categorization for describing students' conceptual model of the environment. For example, in Loughland, Reid, and Petocz's study (2002), students viewed the environment in six distinct ways: (1) the environment as a place; (2) the environment as a place containing living things; (3) the environment as a place containing both living things and people; (4) the environment does something for people; (5) people are part of the environment and are responsible for it; and (6) people and the environment are in a mutually sustaining relationship. The former three categories apparently were object-focused, while the rest focused on the relationship between people and the environment. Littledyke (2004) found that older students tended to view the environment as the world at large while the younger students only thought of what their teachers asked them to do relating to environment protection, such as no littering. Alerby (2000) analyzed student drawings to explore their thoughts about the environment. When thinking about the word environment, some students focused on either the good or the bad world, while some emphasized the dialectics between the good and the bad world, as well as the actions to protect the environment. Shepardson *et al.* (2007) used an environments task to gather students' drawings and written responses to a set of photographs. They identified 12 emergent categories that reflect four mental models of students' conceptions of the environment, including: (1) a natural place where living things live; (2) a place that supports life; (3) a place impacted or modified by human activity or intervention; and (4) a place where living things and humans live.

Several earlier studies aimed to explore students' thoughts about or perceptions of the natural world in contrast to a constructed environment. Van den Born, Lenders, De Groot, & Huijsmani (2001) explored the visions of nature held by the general public (adults between ages 35 and 55) in Europe and the USA. The investigation dimensions included images of nature, images of relationship, and values of nature. The result was positive in that most participants had a strong nature-friendly viewpoint. In a review article, Rickinson (2001) pointed out that young students tended to perceive nature as a non-human environment and held a belief of humans being above and apart from the natural world. It was suggested by Schultz (2000), that concern for environmental problems is closely related to the degree to which people view themselves as part of the natural world. These young students in western countries seemed to maintain "egoistic" concerns for the environment.

Traditional Eastern concepts of nature are described as an emphasis on harmony and oneness of human-nature relationship. However, a comparative study reported in Kellert's book (1996) revealed that neither East nor West possesses an essentially preferable conception of the natural world. In the study, Kellert concluded that the Japanese people's emotional affinity for nature often lacks an ecological understanding so that they showed reluctance to actively participate in protecting wildlife and the environment. Another comparative study (Boyes, *et al.*, 2007) concerning students' conceptions of the quality of the environment showed a distinct result. The non-western (Hong Kong) students were more concerned about air pollution than students in Australia and the UK. From these findings, it can be concluded that experiences, needs and desires, and socio-cultural contexts mediate peoples' perceived values of the environment (Zube, 1987). Peoples' views of the environment could be sometimes fragmentary and fluid depending on what situations they encounter.

Research evidence has shown that people with different backgrounds and life experiences have differing views of the environment. These views may not only reflect how people define and interpret their surroundings, but also affect how they think and make judgments and decisions on the environmental issues. However, few of the investigations on how people conceptualize the environment (nature) were conducted in nonwestern culture (Liu & Lederman, 2007). Interpretations of the relationship between humans and the natural world could differ between cultures at large (e.g., West and East), among groups (e.g., families, organizations), and even on an individual level. In my previous study (Liu and Lederman, 2007), a series of open-ended questions were designed to unearth the differences among academic disciplines in Taiwan. An emergent interpretive framework was developed to describe college students' conceptualization of the natural world from two dimensions, including descriptions about nature and perceptions of human-nature relationships. Finally, there were four summative categories defined for the purpose of comparing with other constructs. From the results regarding environmental worldviews, I found it worthwhile to follow up with further research on this topic because we had very little knowledge about the views held by people of different generations. Therefore, this article reports an empirical study I recently conducted to investigate adolescents' conceptions of the environment. An instrument and analysis framework that is built on the categorization in the literature is designed to explore general views of the environment, ideas about the relationship between humans and the environment, and images about the ideal environment.

Methodology

This is a cross sectional study, involving students of different ages. By this design, differences among groups were detected to understand whether schooling experiences influence students' views of the environment. The research subjects included 81 upper elementary, 88 junior and 130 senior high school students in southern Taiwan. Students of each grade level were selected from one intact class of schools in urban, rural, and remote (indigenous village) areas. The sampled urban area is an industrial city with a well-developed transportation network and contains a multicultural population. For people in the rural area, aquaculture and agriculture are main economic activities. Most people in this area practice folk religion. The rural area schools are located at an indigenous village in the mountainous area with an elevation above 700 meters. Students from these selected schools were invited to participate in this investigation.

The questionnaire contained a series of open-ended questions and a drawing task. The first question was designed by referring to Alerby's study (2000): "What do you think about when you hear the word environment?" It was followed by a request for a drawing that represents the image that comes to mind. Analysis of drawings and written responses to this question attempted to discover the meaning of students' thoughts or descriptions about the environment. In the second question, students were asked to define the central character of the environment. Analysis of the answers to this question was to confirm the drawings and responses from the first one so that students' ideas about what constitutes the environment could be identified. The third question was to directly elicit students' views about the relationship between humans and the environment. In order to encourage students to think about human development and environmental sustainability, the fourth question was formulated as, "If there is one thing that we can do to improve humans' lives but it could destroy the environment; assuming that you have the decision-making power, what would you do? Why would you decide to do this?" After being challenged by the conflict, students engaged in contemplating the ideal environment for them in preparation for the final question. Students' images of the environment were then compared to their ideal.

Since the questions were designed in a coherent and sequential manner, responses to the questions were analyzed in three dimensions: conceptions of the environment, descriptions of the relationship between humans and the environment, and images about the ideal environment. The categories that emerged consisted of a variety of thoughts in association with these three dimensions. Definitions and illustrative quotes for the categories were then developed by three experts to check validity. Moreover, after completing analysis of students' written responses to the questionnaire, 24 students of each grade level were purposefully selected for interviews. The purpose of interviews was specifically to validate researchers' interpretations of written responses, and to further collect students' oral comments about their thoughts in relation to the drawings.

Conceptions of the Environment

Analysis of the students' drawings and responses to the first two questions revealed four qualitatively different conceptions. These conceptions are reported as categories: 1) the earth and universe, 2) a place where plants/animals live, 3) a natural place where people live, and 4) a human-made place where people live. The key words and features associated with each category are described in Table 1 and discussed in the following paragraphs.

Table 1: Categorization of students' conceptions of the environment

Category	Key words / phrases	Example drawing
The earth & universe	- everything on the earth - the great nature on the earth - the earth and space outside of the earth, the whole universe	
A place where plants /animals live	- nature, forest, animals and plants - a habitat for living things - beautiful mountains and rivers, fresh air, clean and unspoiled place - full of living beings	

A natural place where people live	- the arcadia - natural landscape and plantation - a natural place without technology and human destruction - a peaceful place where people and living things live together	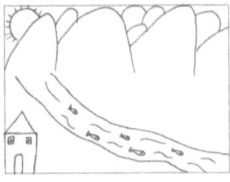
A human-made place	- parks, playgrounds, trails - buildings, roads, cars - pollution, wastes - human exploitation and destruction - city, community, artificial landscape	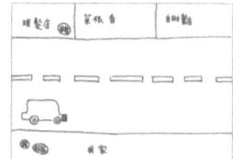

Students whose views were coded in the first category mostly described the environment as the earth, or even the whole universe. This category reflected only 7% of the student responses, among these were five junior high and 11 senior high school students. All of them had drawn pictures containing the earth. When asked who is at the center of the environment, these students provided sophisticated descriptions regarding the complex and dynamic relationships among living and nonliving things on the earth.

The second category, a place where plants and animals live, included 30% of the student responses expressing the environment as a primitive, natural appearance, or a place without human interference. The typical drawing represents the environment as a clean, beautiful and unspoiled area, with sun, clouds, mountains, rivers, trees, birds and fishes. Many of these students held a view that the world would be better without human beings. In addition, none of them considered humans as the center of the environment. The following quote from a senior high student is a typical one.

> "Human beings are not the leading characters of the environment [in my drawing]. If other living and nonliving things on earth become extinct how can people stay alive? Compared with other life forms, humankind is not so important to the earth, because without humans plants and animals may live even better."

Responses in the third category expressed a preference for the natural scenery similar to those in the second category, but placed more emphasis on the role of humankind. This category accounted for 26% of the student responses. It is noted that students who live in the urban area used the drawings and written responses to reflect their images of the ideal environment only to a certain extent. The descriptions of the environment provided by those students living in the remote area, however, revealed their everyday life experiences. The fourth category differs from the third one in that it focused on human constructed environment. This category included 33% of the responses. There were two different orientations in their responses: a neutral and a negative view. The urban younger students tended to neutrally describe their living environment. As seen in the drawing presented in Table 1, this student has labeled his home and some stores in the neighborhood. Many other students were impressed by parks, playgrounds, and recreation facilities in the community, as revealed in their drawings. A portion of students held a more pessimistic view about the status of their living environment by depicting how people destruct and pollute the environment.

Based on the definition of each category, the student responses were coded and then summarized into frequency distribution in order to seek differences among groups with different ages and community settings. The chi-square test was used to examine the significance of the difference. Results showed significant differences among students of different grade levels (Figure 1, next page). The asterisks in the following figures represent significant differences based on the chi-square test results.

Younger students were more likely to describe the scene they had experiences with in their daily life ($\chi2=14.25$, $p<.001$). None of elementary students thought of the environment in a holistic way, with a statement including the earth system ($\chi2=7.08$, $p<.05$). Similarly, fewer elementary students considered the environment as merely a place for animals and plants ($\chi2=7.24$, $p<.05$).

Comparisons among different community settings revealed significant differences as well (Figure 2, next page). More students in urban settings held the fourth category's view, describing a human constructed environment in accordance with their living area ($\chi2=37.47$, $p<.001$). Conversely, students from the remote area tended to describe the environment as a place where animals and plants live ($\chi2=13.08$, $p<.01$). Consistent with previous findings (Wals, 1994; Littledyke, 2004), these results imply that views of the environment are affected by cognitive development and life experience factors. Students' drawings and written responses apparently reflected their images of the place where they live and have had experiences with.

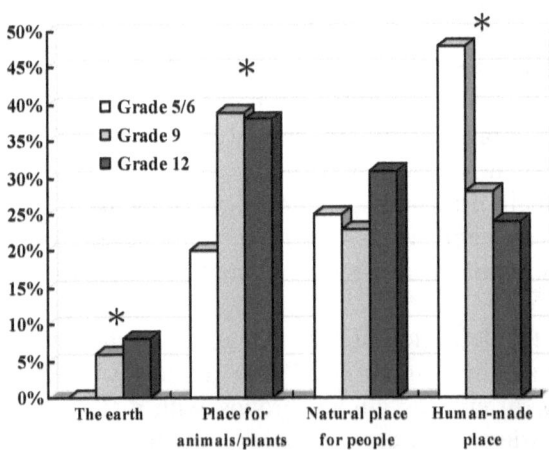

Figure 1. Comparisons of the conceptions of the environment among different grade levels

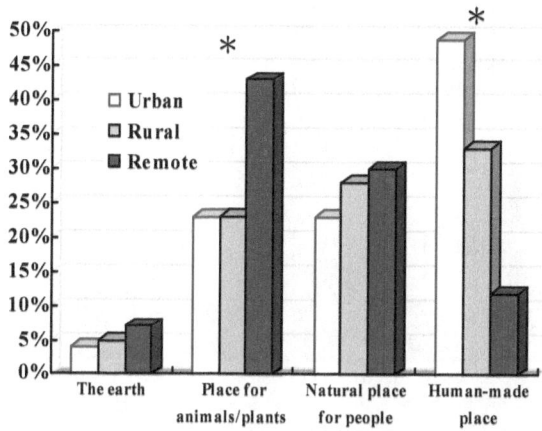

Figure 2. Comparisons of the conceptions of the environment among different community settings

Views of Relationship Between Humans and Environment

Analysis of this dimension mainly focused on the responses to the third and fourth open-ended questions. Loughland *et al.* (2002) suggested that when asked to discuss the relational aspects of humans and the environment, students would be encouraged to gain more integrated conceptions of the environment. The design of these question items was meant to elicit students' comprehensive thinking about the environment. Five categories were emerged: 1) Human beings need the environment and its resources for survival (H→E); 2) The environment needs humans' protection (E→H); 3) Humans and the environment are in a mutually sustaining relationship (H↔E); 4) The environment does not need human beings (E×H); 5) Human beings destroy the environment. Analysis showed significant differences in the relational perspectives of human-environment among grade levels. Frequency distributions are presented in Figure 3 (next page). However, there were no differences among community settings.

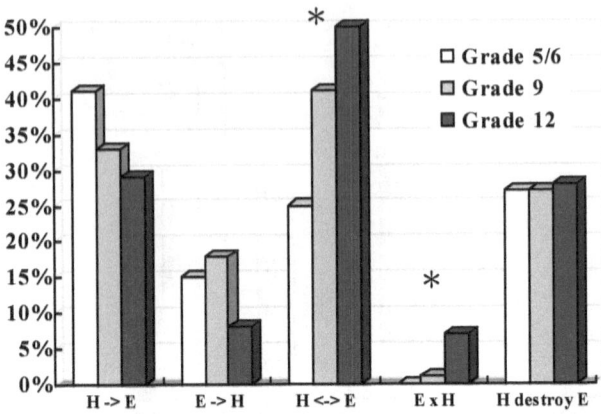

Figure 3. Comparisons of student views of human-environment relationships among different grade levels

The majority of students (40%) indicated a mutual dependence between humans and the environment. Among them, more senior high students than elementary students held this type of view ($\chi 2=13.28$, $p<.001$). Students' views under this category emphasized that people need environmental resources for survival and, in the meantime, the sustainability of the environment needs people's efforts of protection. Only ten high school students (one junior and nine seniors) held a distinct view that humans need the environment but the environment does not necessarily need humans. In addition, when thinking of the reality of human-environment relationship, many students (29%) mentioned that "humans are the main cause of environmental destruction," which is grouped under the fifth category.

In the previous study (Liu & Lederman, 2007), research investigated college students' views of nature by asking interviewees to discuss the human-nature relationships. This study found a similar pattern in that students may give these two objects, (i.e., humans and environment) a different priority and thus describe the relationship from different ontological and philosophical means.

Images of Ideal Environment

The fifth question was designed specifically to ask students about images of an ideal environment where they wish to live. The student responses were coded into five categories, listing in terms of prevalence: 1) a clean, quiet and unspoiled place, 2) a place with natural scenery, 3) a status of harmony, 4) offering people a convenient life, and 5) a rustic village. As Figure 4 (next page) suggests, half of the Taiwanese adolescents in this study expected either a pristine environment or a natural place when they depicted their ideal environment. Nearly one-fifth of the students mentioned the harmonious relationship between self and others (including people and environment). Few students endorsed the view of having a convenient life or living in rustic village. It is evident that the majority of students held an Arcadian image, reflecting their desire to closely connect with nature and to seek peaceful harmony of human-nature interplay.

Comparisons of students' views among grade levels and community settings show that only the views under the category of "a convenient life" were significantly different among urban, rural, and remote areas ($\chi 2=11.64$, $p<.01$). More urban students considered the better environment as making people's life more convenient, having advanced transportation systems and public facilities, and developing good environmental protection technology. In contrast, only one student from the remote area conveyed this view.

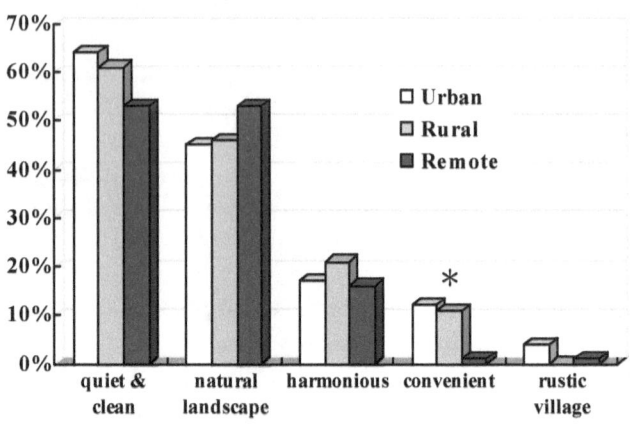

Figure 4. Images of ideal environment: comparisons among different community settings

Statistical analyses were devoted to searching patterns among views of the environment, human-environment relationship, and ideal environment. Only two noticeable patterns were found. First, students who discussed a mutual relationship between humans and the environment and held a view that humans destroy the environment, were more likely to consider the ideal environment as a harmonious status (B=.86 and 1.24, p<.05, based on logistic regression). The second pattern is that students who chose a clean and quiet place as the ideal environment were less likely to state that the environment does not need peoples' protection (B=-1.70, p<.05). The findings imply some consistencies between students' descriptions of the human-environment relationship and what can be viewed as an ideal environment for them. However, no correlations were found between conceptions of the environment and images about the ideal environment. This finding reflects the discrepancy of students' views about what the living environment should be (idealistic) and what the environment actually is (realistic).

Implications to Education

The value of researching students' thinking about the environment is to assist in the development of educational programs. Findings from this study reveal the variations of how students interpret the environment and how they perceive a better environment, including the relationship that human beings should build with the environment. As Loughland *et al.* (2002) suggested a more meaningful approach to environmental science education is one that a student's own experience with the environment is explored and then challenged. By understanding students' conceptions of the environment, educators are able to adapt the environmental curricula to the social and physical context of the school community (Wals, 1994). The research instrument contains a set of coherent question items aimed to elicit respondents' conceptions of the environment from different directions, namely, the realistic, relational, and idealistic perspectives. Each assessed perspective is relevant to the design of experiential learning in environmental education. With careful design and implementation, the questionnaire may be used as part of the instructional materials and assessments, as well as to motivate students towards concern about the environment and environmental issues. Student preconceptions of the living environment provide a starting point for teachers to design learning activities that lead to an increase in "biospheric" concerns (Schultz, 2000), perceiving connections between self and the environment and valuing all living things.

In a reflexive orientation of an environmental education program (Gauthier, Guilbert, & Pelletier, 1997), both teacher and students participate in environmental problem solving. Although a consensus may not be reached, all actors should at least communicate their own conceptual model of the environment with each other. In the learning activities, students should have the opportunities to reflect upon their ideas of the environment. They should also recognize the variations in views among people and, "integrate new experiences and concepts into their mental model" (Shepardson, *et al.*, 2007, p.344). Educators should take students' environmental ideas into account when implementing science and environmental teaching.

Similar to Shepardson *et al.*'s findings (2007), this study revealed that students in urban areas were more likely to view the environment as a place impacted or modified by humans. Learning activities could be designed to direct students to think about how human behaviors impact the environment. In addition to enhancing students' understandings about environmental issues, instructional activities should provide opportunities for students to reflect on their images of an ideal environment and to discuss their role and responsibility to the environment. As Shepardson *et al.* (2007) argued, "Students must learn about the environment before learning about the environmental issue because this places the issue in the context of the environment" (p. 344).

Results of this study further suggest that for those students from urban settings, lack of nature experiences limited their views of the environment. There is an increasing problem of diminished human contact with nature, a phenomenon that is called "Nature-deficit disorder" (Louv, 2006). Young people living in the urban areas have fewer chances than their rural counterparts to access the natural world. A new and recent educational movement in the United States is the 'No Child Left Inside' Act, giving children opportunities to have outdoor experiences and to learn about nature. Results of this investigation also underscore the importance of outdoor learning and field studies, especially for students living in urban areas. There is convincing evidence that childhood nature experiences are associated with nature-friendly behavior in adults (e.g., Van den Born, *et al.*, 2001). Recently, many nature centers and environmental learning centers have been supported by governmental or private sectors in Taiwan in order to achieve the goal of having children experience nature. Furthermore, informal environmental learning experiences, such as attending educational programs in nature centers, have been proven to advance science learning (Knapp & Barrie, 2001).

Data from students in the remote area indeed raised the researcher's attention. These students are mostly from indigenous tribes. They often viewed the living environment as full of animals, plants, and natural scenery and appeared to be satisfied with this kind of environment. It seems that these indigenous adolescents value the place where they live. Research on the human-place bond, proposed by environmental psychologists (e.g., Gustafson, 2001; Manzo, 2003), suggests that everyday experiences of place are related to the conceptualization of place. People's attachments to a specific place, such as a local community environment or a natural setting, reflect their evaluation and value judgment of the condition associated with that place (Kyle, et al., 2004). These indigenous students possibly value the natural environment and their residential community. Their commitments to environmental conservation can be anticipated as long as their positive experiences of the environment are sustained. This study's opinion regarding curriculum design is in agreement with multicultural education, suggesting that the development of curricula for different cultural groups should include materials from the local environment and also be incorporated with traditional practices.

Students' conceptions of the environment could be shaped by several factors: learning and life experiences, community settings, gender, nature experiences, age, and cognitive development (Rickinson, 2001). This study only explored a few of these factors. Quantitative statistics can explore the relationship between students' conception of the environment and demographic variables such as age and school location. This study intended to learn more about how students' environmental perspectives change over different ages by a cross sectional design. However, a more qualitative oriented, longitudinal study tracking students' views over time may be needed to gain insight into the influence of learning experiences and cognitive development.

As an exploratory study, this research is the first attempt at gaining the phenomenological understanding of perspectives of different aged students in Taiwan. The open-ended questionnaire design gathered responses realistic to people's thinking. Future research can utilize the sophisticated description of analysis framework to expand investigation to a wider range of population, including different generations and cultural settings. Comparisons among populations are needed (Payne, 1998). As science and environmental educators, we should also reflect on our own perspectives of the environment and compare with our students' views in order to build a better communication platform for discussing environmental issues.

References

Alerby, E. (2000). A way of visualizing children's and young people's thoughts about the environment: A study of drawings. *Environmental Education Research, 6*(3), 205-222.

Astran, S., & Medin, D. L. (2008). *The native mind and the cultural construction of nature.* Cambridge, MA: MIT Press.

Astran, S., Medin, D. L., & Ross, N. O. (2005). The cultural mind: Environmental decision making and cultural modeling within and across populations. *Psychological Review, 112*(4), 744-776.

Boyes, E., Myers, G., Skamp, K., Stanisstreet, M., & Yeung, S. (2007). Air quality: A comparison of students' conceptions and attitudes across the continents. *Compare: A Journal of Comparative Education, 37*(4), 425-445.

Daugstad, K., Rønningen, K., & Skar, B. (2006). Agriculture as an upholder of cultural heritage? Conceptualizations and value judgments - A Norwegian perspective in international context. *Journal of Rural Studies, 22*, 67-81.

Gauthier, B., Guilbert, L., & Pelletier, M. L. (1997). Soft systems methodology and problem framing: Development of an environmental problem solving model respecting a new emergent reflexive paradigm. *Canadian Journal of Environmental Education, 2*, 163-182.

Gustafson, P. (2001). Meanings of place: Everyday experience and theoretical conceptualization. *Journal of Environmental Psychology, 21*, 5-16.

Kellert, S. R. (1996). *The value of life: Biological diversity and human society.* Washington, D.C.: Island Press.

Knapp, D., & Barrie, E. (2001). Content evaluation of an environmental science field trip. *Journal of Science Education and Technology, 10*(4), 351-357.

Kyle, G., Graefe, A., Manning, R., & Bacon, J. (2004). Effects of place attachment on users' perceptions of social and environmental conditions in a natural setting. *Journal of Environmental Psychology, 24*, 213-225.

Lijmbach, S., Margadant-Van Arcken, M., Van Koppen, C. S. A., & Wals, A. E. J. (2002). "Your view of nature is not mine!": Learning about pluralism in the classroom. *Environmental Education Research, 8*(2), 121-135.

Littledyke, M. (2004). Primary children's views on science and environmental issues: Examples of environmental cognitive and moral development. *Environmental Education Research, 10*(2), 217-235.

Liu, S.-Y., & Lederman, N. G. (2007). Exploring prospective teachers' worldviews and conceptions of nature of science. *International Journal of Science Education, 29*(10), 1281–1307.

Loughland, T., Reid, A., & Petocz, P. (2002). Young people's conceptions of environment: A phenomenographic analysis. *Environmental Education Research, 8*(2), 187-197.

Louv, R. (2006). *Last child in the woods: Saving our children from nature deficit disorder.* Chapel Hill, N.C.: Algonquin Books of Chapel Hill.

Manzo, L. C. (2003). Beyond house and haven: toward a revisioning of emotional relationships with places. *Journal of Environmental Psychology, 23*, 47-61.

Mueller, S. T. & Veinott, E. S. (2008). Cultural mixture modeling: Identifying cultural consensus (and disagreement) using finite mixture modeling. In B. C. Love, K. McRae, & V. M. Sloutsky (Eds.), *Proceedings of the 30th Annual Conference of the Cognitive Science Society* (pp. 64-70). Austin, TX: Cognitive Science Society.

Payne, P. (1998). Childrens' conceptions of nature, *Australian Journal of Environmental Education*, 14, pp. 19- 26.

Rickinson, M. (2001). Learners and learning in environmental education: A critical review of the evidence. *Environmental Education Research, 7*(3), 207.

Robertson, A. (1993). Eliciting students' understandings: Necessary steps in environmental education. *Australian Journal of Environmental Education, 9*, 95–114.

Schultz, P. W. (2000). Empathizing with nature: The effects of perspective taking on concern for environmental issues. *Journal of Social Issues, 56*(3), 391-406.

Shepardson, D. P., Wee, B., Priddy, M., & Harbor, J. (2007). Students' mental models of the environment. *Journal of Research in Science Teaching, 44*(2), 327-348.

Shutters, S. T. & Cutts, B. B. (2008). A simulation model of cultural consensus and persistent conflict. In V.S. Subrahamanian & A. Kruglanski (eds.), *Proceedings of the Second International Conference on Computational Cultural Dynamics* (pp. 71-78). Menlo Park, California: AAAI Press.

Van den Born, R. J. G., Lenders, R. H. J., De Groot, W. T., & Huijsmani, E. (2001). The new biophilia: An exploration of visions of nature in Western countries. *Environmental Conservation, 28*(1), 65–75.

Wals, A.E.J. (1994). Nobody planted it, it just grew! Young adolescents' perceptions and experiences of nature in the context of urban environmental education. *Children's Environments*, 11(3), pp. 177-193.

Zube, E. H. (1987). Perceived land use patterns and landscape values. *Landscape Ecology, 1*(1), 37-45.

www.ingramcontent.com/pod-product-compliance
Lightning Source LLC
Chambersburg PA
CBHW031544300426
44111CB00006BA/167